INSIDE CAMBODIAN INSURGENCY

Military Strategy and Operational Art
Edited by Professor Howard M. Hensel, Air War College, USA

The Ashgate Series on Military Strategy and Operational Art analyzes and assesses the synergistic interrelationship between joint and combined military operations, national military strategy, grand strategy, and national political objectives in peacetime, as well as during periods of armed conflict. In doing so, the series highlights how various patterns of civil–military relations, as well as styles of political and military leadership influence the outcome of armed conflicts. In addition, the series highlights both the advantages and challenges associated with the joint and combined use of military forces involved in humanitarian relief, nation building, and peacekeeping operations, as well as across the spectrum of conflict extending from limited conflicts fought for limited political objectives to total war fought for unlimited objectives. Finally, the series highlights the complexity and challenges associated with insurgency and counter-insurgency operations, as well as conventional operations and operations involving the possible use of weapons of mass destruction.

Also in this series:

Air Power in UN Operations
Wings for Peace
Edited by A. Walter Dorn
ISBN 978 1 4724 3546 0

Understanding Civil-Military Interaction
Lessons Learned from the Norwegian Model
Gunhild Hoogensen Gjørv
ISBN 978 1 4094 4966 9

Clausewitz's Timeless Trinity
A Framework For Modern War
ISBN 978 1 4094 4287 5

British Generals in Blair's Wars
Edited by Jonathan Bailey, Richard Iron and Hew Strachan
ISBN 978 1 4094 3735 2

Britain and the War on Terror
Policy, Strategy and Operations
Warren Chin
ISBN 978 0 7546 7780 2

Inside Cambodian Insurgency
A Sociological Perspective on Civil Wars and Conflict

DANIEL BULTMANN

Humboldt University Berlin, Germany

ASHGATE

Published by
Ashgate Publishing Limited
Wey Court East
Union Road
Farnham
Surrey, GU9 7PT
England

Ashgate Publishing Company
110 Cherry Street
Suite 3-1
Burlington
VT 05401-3818
USA

www.ashgate.com

British Library Cataloguing in Publication Data
A catalogue record for this book is available from the British Library

The Library of Congress has cataloged the printed edition as follows:
Bultmann, Daniel.
 Inside Cambodian insurgency : a sociological perspective on civil wars and conflict / by Daniel Bultmann.
  pages cm
 Includes bibliographical references and index.
 ISBN 978-1-4724-4305-2 (hardback : alk. paper)—ISBN 978-1-4724-4306-9 (ebook)—ISBN 978-1-4724-4307-6 (epub)
 1. Cambodia—History—Civil War, 1970–1975. 2. Social conflict—Cambodia—History—20th century. I. Title.
 DS554.8.B85 2015
 959.604'2—dc23

2014028396

ISBN: 9781472443052 (hbk)
ISBN: 9781472443069 (ebk – PDF)
ISBN: 9781472443076 (ebk – ePUB)

Printed in the United Kingdom by Henry Ling Limited,
at the Dorset Press, Dorchester, DT1 1HD

Contents

List of Figures

List of Tables

List of Abbreviations

ANS	Armée Nationale Sihanoukiste
CGDK	Coalition Government of Democratic Kampuchea
CPK	Communist Party of Kampuchea
ECCC	Extraordinary Chambers in the Cambodian Courts
FANK	Forces Armées Nationales Khmères
Funcinpec	Front Uni National pour un Cambodge Indépendant, Neutre, Pacifique, et Coopératif
KNUFNS	Kampuchean National United Front for National Salvation
KPNLF	Khmer People's National Liberation Front
KPNLAF	Khmer People's National Liberation Armed Forces
MOULINAKA	Mouvement de Liberation National du Kampuchea
NADK	National Army of Democratic Kampuchea
PDK	Party of Democratic Kampuchea
PRK	People's Republic of Kampuchea
RCAF	Royal Cambodian Armed Forces
UNBRO	United Nations Border Relief Operation
UNICEF	United Nations Children's Fund
UNHCR	United Nations High Commissioner for Refugees
UNTAC	United Nations Transitional Authority in Cambodia

Frontispiece Drawing by a Khmer Rouge cadre during a political lesson. Source: Documentation Center of Cambodia. The caption reads: "Combatants protecting the border."

Acknowledgments

Without the willingness of countless respondents and informants to share their lifecourse, experiences, thoughts, time, and sometimes their personal networks, this project would not have been possible. Even though our talks touched on highly sensitive, emotional, and ethically precarious topics, most respondents were eager to answer each question and to delve deep into the inner (dys)functionings of their units. I am very thankful for thoughtful guidance and restless support from Prof. Dr. Boike Rehbein at Humboldt University, who has accompanied and 'educated' me since I started my research on Cambodian conflicts several years ago. Likewise, Prof. Dr. Klaus Schlichte from the Institute of Intercultural and International Studies (InIIS) at the University of Bremen became an indispensible companion and supervisor to my work. I am very proud and thankful that Prof. Dr. Howard Hensel puts so much trust in this book and its sociological perspective for his series on *Military Strategies and Operational Arts*. Moreover, without the generous funding of the Elsa Neumann Foundation as well as the provision of travel funds by the German Academic Exchange Service (DAAD), this project would have been impossible from the start.

During my fieldwork I met many committed supporters who undertook various efforts and provided much needed support to facilitate this study. Sok Udom Deth has accompanied me as a good friend and advisor for several years now. Likewise, my research assistant Koy-Try Teng, who had to face many difficult if not sometimes deeply distressing situations throughout this project, became not only a great interviewer but also a friend. I will always have good memories of our travels together with Rik Chor, who spent months with us in search of additional respondents. Rik Chor's patience, commitment, and knowledge in the field are beyond comparison. I do hope to get the chance to meet him and Mr. Koy at some point again in the near future. Special thanks go to Kem Sos, who opened various doors for us and silently but effectively paved our way through Cambodia.

I also would like to thank Sothy Eng for wonderful times in Cambodia; Astrid Noréen-Nilsson for the (although in the end fruitless) support in contacting 'the commander'; Kosal Phat for guiding us in troubled times at Anlong Veng; Sary Seng for introducing us to the 'camp village'; Abdul Gaffar Peang-Meth for his dedication from overseas; Robert Karniol, Suon Khieu, and Kong Thann for kindly sharing their network, time, and knowledge; Sim Sorya for advice on the ministry; Robert Carmichael, Laura McGrew, Lindsey French, and Mitch Carleson for sharing some network connections and experiences; Benjamin Schwalb for triggering my thoughts on the inner functionings of armed groups; Hi-Khan Truong for 'professionalizing' some of the graphs; the *Documentation Center of Cambodia* (DC-Cam), most notably Youk Chhang, Dany Long, and Kok-Thay Eng, for expertise and documents; Kenneth Conboy for his valuable comments; and last but not least Annie McWhertor and Christopher O'Mahony for proofreading the draft of this text. I am deeply indebted to my wife, Johanna, my little daughter, Mathilda, and my brother, Christof, whose share in facilitating this project go way beyond the single line of appreciation recorded here.

Introduction

The chain of command was well-rehearsed and functioned smoothly, enabling tight discipline to be maintained on fire control during engagements. (Guest, 1985, p. 598)

While reports like this on the military discipline of Khmer Rouge soldiers were usually highly reverential, their counterpart within the tripartite *Coalition Government of Democratic Kampuchea* (CGDK), the *Khmer People's National Liberation Armed Forces* (KPNLAF), received mostly rather negative quotes. Even within internal CIA reports on their Cold War clients fighting against Soviet-backed Vietnamese occupation along the Cambodian border, the US patrons did not waste much time sugarcoating their view: "They are hindered by a number of long-standing deficiencies in command and control, logistics, leadership talent, and cohesion" (CIA, 1989, p. 4). The third faction within the coalition, in turn, the royalist *Front Uni National pour un Cambodge Indépendant, Neutre, Pacifique, et Coopératif* (Funcinpec), was highly recognized in the political arena due to their leader: former King Norodom Sihanouk. However, on the battlefield, its military wing *Armée Nationale Sihanoukiste* (ANS) went largely unnoticed despite some unexpectedly well-performed defense operations towards the end of Vietnamese occupation. While disciplinary structures are the main instruments that constitute power in armed groups, they are seldom studied, despite an increased interest in organizational modes within armed groups in recent research on civil wars. The central question of this study, therefore, is how power is constituted and secured. How are different types of disciplinary structures within armed groups constituted and reproduced? Which power institutions and practices are used to make soldiers obedient to command, by whom and why? And how do soldiers themselves structure and deal with discipline?

While interviewing commanders for this project, the wide range of possible power practices used to constitute discipline became obvious. All factions of the CGDK showed different strategies for ensuring that their soldiers were disciplined, loyal, and responsive to orders. Some, for example, punished them using transformative techniques such as re-education while others barely punished them at all. While some trained their soldiers in close-knit and highly formulaic combat simulations, others simply handed them weapons without any sort of preparation, instead engaging in 'management of flows'. There is a wide range of possible disciplinary practice within the field, and sometimes the lack thereof was the most interesting. A group's political orientation and ability to recruit soldiers are not sufficient grounds for explaining these variations, for the differences seem to be similar across different groups and, conversely, may differ sharply between commanders who are active within the same group. The basic assumption of the sociological approach using Pierre Bourdieu's field theory, however, is that these variations do not emerge at random but contain logic and regularities, thereby offering a better insight into the micro-foundations of non-state armed groups.

Instead of looking at random practices, the hypothesis of the field-theoretical approach being proposed here is that classificatory discourses among different 'habitus groups' constituting the field of military resistance structure the group's power practices.[1] First and foremost, those classificatory discourses encompass the attitudes that military commanders have regarding the 'nature' of their soldiers and which ideals of skillful leadership and what it means to be a soldier they apply in order to govern their power practices. Symbolic classifications among leading 'habitus groups' thereby serve as patterns for dominant practices and institutions such as training, indoctrination, and system of rewards and punishments, as well as combat control. Furthermore, prevailing classifications among leading 'habitus groups' correspond with the different Foucaultian power rationales that dominate each habitus group. Thereby Foucault's analysis of power forces is reinterpreted as or becomes embedded within a relational space of social forces. Power and discursive relations are social relations within the social structure of a field. Each habitus group has a distinctive classificatory discourse with distinctive patterns of power practice.

Each habitus group within the leadership has a resource of which it disposes most and which the agents try to promote in the field. This 'symbolic resource', such as having a comparatively high level of education or surviving many battles, is the result of an agent's social background and lifecourse. Additionally, it structures the classificatory discourse and henceforth the application of power practices within the group. Thus, discourse and practices reflect an attempt to increase the value of the group, its main symbolic resource, and thereby its position in the field. However, analyzing the habitus of agents in their diachronic formation is imperative as well. The study maintains that differences in habitus formation can only be understood by taking an agent's lifecourse into account. Pierre Bourdieu himself largely overemphasizes the synchronic formation of an agent's habitus, in which an agent simply incorporates or mirrors his structural position over the course of time. Thereby an agent becomes what he already is in terms of his structural position in a nation's social structure. Instead of fixed positions on a flat social space, the current study argues for a more dynamic approach, using Pierre Bourdieu's own concept of a social field.

---

[1]   At times I am using the term 'resistance' although it might carry some ideological connotations. One reason for my decision to do so is that respondents themselves relate to the field as 'the resistance'. It is a term used to glorify their efforts, and it is clear that in many cases it was not pure 'resistance' that motivated people to join these non-state armed groups. But any other term used in the text, such as insurgency, is slightly off track as well. Moreover, most terms were coined for armed groups in African conflicts, in which economic agendas and greed are suspected to motivate warlords to carry on with 'fighting' only to sustain their 'businesses'. Framing their agenda as 'resistance' in order to connect local motives with a master cleavage does, of course, arouse suspicion regarding its merely legitimizing function. Any of these terms carries either the connotation of grievance or greed-led wars and often tells something about the political and theoretical positioning of the scholar using it, too.

The study yields several insights into the formation of armed groups, their power relations, and their social structure. Two corresponding results are of particular importance. First, to a large extent, the field of insurgency reproduces social differentiations that existed prior to its own formation. Many positions in the field are homologous with older hierarchies, e.g., with an old military and political elite forming its upper ranks, their patrimonial clients within the mid-range leadership, and a displaced peasantry within the rank and file. This shows that social differentiations survive even massive societal changes, with political networks being reconstituted and militarized. Second, however, at the same time, a field cannot be explained solely by reference to a prior social structure being reproduced; its own historical and symbolical formation must also be considered. Each field has a different history and symbolic universe that values resources differently. Therefore, some groups are able to rise in status, such as due to an increased symbolical value of combat experience within a society at war, and an insurgency in particular. This means that the study maintains that practices within armed groups are socially differentiated. Recent research shifted towards the micro-foundations of armed groups' practices. Within influential entities, the patterns of (violent) practices and organizational modes of armed groups were analyzed as either spatially (Kalyvas, 2006) or economically (Weinstein, 2007) differentiated. In contrast to these studies, however, the current study maintains the social differentiation of organizational modes and practices.

While most recent studies on civil wars tried to determine the logic behind the use of violent practices against civilians, more specifically why groups manhandle civilians, this study, by contrast, analyzes the power practices and disciplinary structures within armed groups. It does so for two reasons. First, interviewing combatants and commanders regarding the violence against civilians does not, in my view, yield many accurate accounts by the respondents, due to the sensitivity of the topic. No one talks at ease about a massacre, rape, and other such violent practices. This, in my view, tackles the veracity of studies that use questionnaires for quantitative analysis. Reluctance to directly address violent practices against civilians makes any study on armed groups complicated if not impossible. However, this is also the case for qualitative studies, which end up without reliable data when trying to make ex-combatants talk about things that would be categorized as 'war crimes' by most observers. Especially in cases where there is a war crimes tribunal (as in the Cambodian case with the *Extraordinary Chambers in the Cambodian Courts* (ECCC) passing judgments on the Khmer Rouge genocide), it is almost impossible to circumvent the reluctance on the side of the respondents. Hence, data may look reliable when gathered and accumulated for quantitative analysis, but they rest upon shaky foundations. The second reason to study power practices and disciplinary structures is that it leads to an examination of the micro-foundations of armed groups as such. Disciplinary practices and institutions are well suited to study how power is practiced and maintained in armed groups. This gives researchers an opportunity to analyze how all belligerents in the field participate in the maintenance of civil war and the existence of resistance forces.

The main method used to gather material was to conduct semi-structured interviews with eighty-six respondents from across each faction and rank. Not every interview is cited in the text, in order to keep the material manageable and clear (sixty interviews are cited in all). 'Semi-structured' means that a guiding open question was posed at the beginning of each of the three topics to be discussed, followed by some more specific inquiries, which can be omitted or altered during the interview depending on its progression and the mood of the respondent. The first part comprised questions pertaining to the respondents' family, lifecourse, upbringing, and education. The second part focused on their sentiments regarding the nature of their leaders or subordinates, as well as their concept of good leadership and soldiers. This section provided respondents with an opportunity to elaborate on the classificatory discourse. The last part of the interview investigated power practices such as military training, indoctrination, and battle organization and control, as well as systems of reward and punishment. Most interviews were conducted in the Khmer language, and some were held in English. The method used to analyze information related to the respondents' lifecourse follows 'habitus hermeneutics' as it was and still is developed by Helmut Bremer, Michael Vester, and Andrea Vester-Lange and is currently applied to the analysis of global inequality (Bremer, 2004; Lange-Vester, 2007; Rehbein, 2008). All interviews were transcribed word-by-word, translated into English, if necessary, and analyzed sequentially.

The study is divided into seven parts. The first chapter situates the study within the field of research on civil war, particularly on micro-politics of non-state armed groups. Much of the older and ongoing research still focuses on the division between greed and grievance practices allegedly characterizing civil wars. The second chapter briefly turns to the theories of habitus formation, social fields, and power. Chapter 3 introduces the structure and history of the field. Chapters 4 to 6 discuss the habitus groups within the field, their respective classificatory discourse, and their distinctive power practices. While Chapter 4 elaborates on the main theory presented in this work with regard to the leadership which defines most of the binding rules within the field, Chapter 5 describes the transmission of practices by the mid-range in command. The strategies that the rank-and-file soldiers used to deal with disciplinary structures are analyzed within sixth chapter. The last chapter systemizes insights gathered from the study for the theory and analysis of insurgent movements and also explains how Pierre Bourdieu's and Michel Foucault's approaches need to be refined.

Chapter 1
Greed and Grievances at War

Most observers of the Cambodian Civil War said that it was a mere proxy war caught in major Cold War competition, in which local agents were more or less pawns in the game (e.g., Reynell, 1989). Therefore, any need to delve deeper into its social and military structures and local politics was seen as a dispensable intellectual exercise. Analysts interpreted almost all small-scale conflicts during the Cold War as sideshows of the struggle between the US and the Soviet Union. The Cambodian case is a seemingly perfect example, in which the US and China formed a coalition to strengthen the warring factions along the border in their war against the Soviet-backed Vietnamese and their 'hegemonic expansion'. Without their military, financial, and political support, the struggle of the resistance certainly would have ended quickly. However, after the end of the Cold War, many of the conflicts went on (such as in Cambodia), and many new conflicts arose around the world. The 1990s saw some of the bloodiest conflicts and a trend towards internal instead of interstate conflicts. The big Cold War could no longer serve as an explanation for those small wars.

A first explanation was that underneath the Cold War conflict, smaller states or even communities made use of the larger framework and military support to wage their ethnically motivated skirmishes. After the breakdown of the Soviet Union, the ethnic dimension (re-)surfaced. Many simply saw a resurgence of nationalist and ethnic movements (Wimmer, 2002; Fearon & Laitin, 2000). In light of conflicts in Africa, the American journalist R. Kaplan even announced a 'coming anarchy' with internal wars appearing to be rather anarchical and increasingly ferocious along ethnical lineages, making them almost impossible to end (2001). And for him, ancient hatreds were also at work in Yugoslavia, which is why he called his book *Balkan Ghosts* – ghosts resurfacing after the end of the Cold War and constantly haunting society (2005). For many observers, ethnic grievances lay underneath the seemingly anarchic surface of these small wars (van Creveld, 1991), a dimension that had been hidden by the Cold War script and now dominated the conflicts arising around the globe. For these authors, the ethnic dimension is deep rooted and thereby practically irresolvable (e.g., Kaplan, 2005). The only option is to let those conflicts cool down, but since they are rooted in a primordial identity, they will resurface again and again. This leaves policy makers with few options at their disposal to end the conflicts – both then and now.

However, more nuanced analyses of conflicts driven by 'communal identity' are those presented by Edward Azar and Ted Gurr, who both developed their ideas during the Cold War, when most scholars saw no need for studying small-scale proxy wars. For Edward Azar, protracted social conflicts are shaped by the struggle of communal groups "for such basic needs as security, recognition and acceptance, fair access to political institutions and economic participation" (1991, 93). In his study, he emphasized need deprivation and governance failure as causes for violent conflict but also the need for multi-factorial approaches in conflict studies and a mapping of all agents, forces, and institutions in the field (1990). Ted Gurr's theory of relative deprivation was another influential and multi-factorial approach among social grievance theorists during the Cold War era, operating with in-group and out-group antagonisms. His core argument, however, is that people rebel because they perceive an inequality in comparison to other groups, resulting in a lower position in society than they themselves believe they are entitled to. Therefore, he simply adds to Azar's theory of deprivation the aspect of comparison between groups, who *perceive* a deprivation *relative* to others. Particularly in situations in which a group faces a stark downturn in social status, these comparisons can instill violent action against a group of 'others' serving as scapegoats – often these groups then formulate and interpret differences to others in ethnic and primordial terms.

New micro-political approaches, however, highlight the gap between seemingly existing needs (Azar) and grievances (Gurr) and their mobilization by the leaders of rebellions (cf. Demmers, 2012, pp. 93–5). Leaders do not simply 'pick up' existing agendas in the public and thereby mobilize for collective action; these agendas are also socially and discursively constructed. On the basis of Erving Goffman's concept of 'framing' (1974), Sidney Tarrow coined the term 'collective action frames' for the construction of power in social movements by 'framing' (1998). Using this concept to explain the construction of an agenda in rebellions, Snow and Benford define a 'frame' as an "interpretative schemata that simplifies and condenses the 'world out there' by *selectively* punctuating and encoding objects, situations, events, experiences, and sequences of actions within one's present or past environment" (1992, p. 136; emphasis added). Snow and Benford sketch cycles of protest in relation to certain master frames inside rebellions, while highlighting the contested nature of these frames over the course of conflict and inside a movement. There is no agenda waiting to be uncovered by the leadership, but these agendas are selectively constructed, evolve over time, and vary between different factions in a movement.

During the late 1990s, a second competing theoretical framework emerged. This new approach depicted itself as a competing framework that highlighted a qualitative change in warfare after the Cold War, which is held responsible for the arising conflicts, especially in the Global South. Many scholars started to see a predominant economic motivation at work in post-Cold War conflicts. Small wars were not fought for communities but for money, which is why looting, robbery, and resource extraction become increasingly widespread in these conflicts (Collier & Hoeffler, 2001; Berdal & Melone, 2000; Elwert, 1999). Or, as Keen (1998) termed it in reference to Clausewitz's famous sentence: War becomes the continuation of economics by other means. The state breaks down and warlords preside over purely profit-oriented markets of violence (e.g., Reno, 1998). More and more scholars, moreover, have studied the global embeddedness and interconnectedness of post-Cold War conflicts that seem to lack a political cause and are increasingly so-called 'network wars', whose agents are spread all over the global sphere (Nordstrom, 2007; Duffield, 2002; Holsti, 1996). Many scholars saw a qualitative change in warfare after the Cold War, leading to the advent of increasingly asymmetrical, globalized, and economy-driven 'new wars'. While 'old wars' were regulated, politically motivated, and symmetrical (two equal armies fight), 'new wars' are unregulated, asymmetrical, increasingly ferociously fought against civilians, and part of a greedy and globalized criminal network (Münkler, 2005; Kaldor, 1999).

In their most radical and clear-cut version, 'new wars' are fought for nothing other than economic resources, with participants seeking personal benefits at the local and the global level and with warlords extracting easily marketable and portable resources such as diamonds and profiting from a globalized economy (e.g., Berdal & Melone, 2000; Collier, 2000). While those 'greedy conflicts' happen in 'broken states' at the periphery of the world system in 'post state societies', their economy, and thereby also their motives, are related to financial centers around the world. Hence, there are no politics in a 'broken state', just naked economic self-interest and globalized exploitation by criminal networks. Stathis Kalyvas summed up the highly biased and stylized distinction between 'new' and 'old' wars in three points:

> 1. Old civil wars were political and fought over collectively articulated, broad, even noble causes, such as social change – often referred to as 'justice'. By contrast, new civil wars are criminal and are motivated by simple private gain – greed and loot.
> 2. At least one side in old civil wars enjoyed popular support; political agents in new civil wars lack any popular basis.
> 3. In old civil wars, acts of violence were controlled and disciplined, especially when committed by rebels; in new civil wars, gratuitous and senseless violence is meted out by undisciplined militias, private armies, and independent warlords for whom winning may not even be an objective. (2001, p. 102)

By opening up the distinction between the politics and economics in wars, these authors from the 'Global North' define a highly problematic distinction between 'their wars as greedy wars' and 'our wars as political wars' (cf. Duffield, 2002). Several scholars, however, have begun to call for a closer and more nuanced examination of the social and political function of the seemingly economy-driven and loot-seeking forms of contemporary warfare such as warlordism (e.g., Demmers, 2012, p. 70; Keen, 2008; Cramer, 2006; King, 2004). The latest studies, moreover, also point towards the political role that refugee and diaspora communities play in increasingly globalized wars, offering a more balanced view that goes beyond a simple dichotomy between politics and the economy (Schlichte, 2012; Falge, 2011; Ballentine & Sherman, 2003; or on cultural globalization in recent warfare: Shepler, 2005).

Certainly, wars are not always fought and prolonged solely for political reasons, as economic interests do play an important role (a good example of a balanced study being Jean & Rufin, 1999). Particularly demobilization and conflict resolution are confronted with the problem that many agents are much too well embedded in the economy of warfare and fear losing the profitable position they have gained over the years if demobilization and reintegration are successful. Their resources, whether economic, social, or cultural, might not matter during peace as much as in times of war because they are rather specific to a field at war. Or, as Keen maintained, "wars very often are not about winning" (2008, p. 15). Greed theories, however, especially as outlined by Paul Collier's classic World Bank study (2000; with Hoeffler, 2001), interpret patterns of violence by propagating these patterns as their own and sole source. Simply put, while indiscriminate violence against communities occurs due to an underlying ethnic and social script, looting and resource extraction (diamonds or other gems in particular) occur within conflicts with an economic script. Within greed approaches, the individual's or group's economic motives preceding membership in a rebellion explain their economic activities within and over the whole course of conflict. This is easily supplemented with a rationalist model of action, in which the agent seeks to increase his personal profit by joining (or staying within) a rebellion. Hence, having many cases of looting or of other proxies of 'economic' factors (such as low education, poverty, many young males, etc.) serves as evidence for hidden greed scripts that underscore the public discourse of grievances, as put forward by the leadership of a rebellion. Many scholars have already pointed towards the shaky evidence when statistical sets are used, in which some acts and factors are unambiguously declared as proxies for 'greed' and others for 'grievance' (Demmers, 2012, p. 103). Also, the simple fact that poverty is not a good predictor of rebellion contradicts all expectations (cf. Goodhand, 2003).

Greed theories literally try to calculate the costs and benefits of participating in rebellion and to find a reasonable threshold for 'rational' participation or a large enough benefit that could account for the huge risks that participants take, such as the risk of dying in battle. Thereby the main agents at war are not soldiers or political leaders but economic resources: Depending on the composition of resources at hand, the rational agent opts in favor of certain practices and participation in armed conflict, or against them. Sometimes this is supplemented with different types of individual and collective profit in order to maintain the general theoretical underpinnings in cases where the model does not work smoothly. While most scholars rejected greed theory as simplistic and flawed, some, such as Collier (2007), tried to save the general impetus of including economic aspects into the study of war. However, many global institutions such as the World Bank and other development organizations still find the 'results' useful, largely because they serve as an explanation for the failure of long-time neoliberal programs in the developing world. With these 'findings' at hand, they could and still can place the blame on local conditions and agents outside of their own political sphere and programmatic reach (cf. Demmers, 2012, pp. 100–115). And, on a more pragmatic level, the idea of economic calculus leaves room for the possibility that there are ways to end a conflict – in contrast to concepts of primordial ethnic cleavages that leave policy makers with empty hands. Thus, the very existence of demobilization agencies, providing incentive packages for ex-combatants to leave the rebellion, becomes legitimate in the first place.

Stathis Kalyvas (2004; 2003), however, was the first to cut the so-far-essential explanatory relationship between motive (greed/grievance) and warfare by relating certain patterns of violence in a conflict to the degree of control an armed group exerts in a certain area and its relationship to the local populace. Certain types of violence are used not because of greedy motives but because of the logic of control over a populace. Hence, warfare has its own dynamic that cannot be deduced and explained solely by reference to the motives preceding it. The type of violence is not related to the motive of agents. Instead, depending on the degree of popular control, an armed group 'opts' either for indiscriminate or selective violence. By maintaining a relationship between different degrees of control and the type of violence being exerted, abusive and non-abusive or selective and indiscriminate acts of violence become *spatially differentiated*. While greed or grievance motives do not simply translate into respective practices for Kalyvas, he still uses a similar rationalist and highly mathematical model to investigate violent practices against the civilian population operating with an anthropology of a *homo oeconomicus* – just with a changed, non-economistic rationality at work. However, Kalyvas's theory is part of a 'micro-political turn' in which formerly analytical blocs of conflict are dissected in detail across time and space (cf. King, 2004). This approach looks at the variation of violence between different stages of conflict or across different villages, cities, or areas and searches for the reasons behind these variances. In this theory, there is no ethnic war but 'indiscriminate violence' against communities that occurs at certain places and stages of warfare.

Kalyvas and Kocher (2007a), for example, show how the Iraqi War can be divided into at least five different conflicts underneath the seemingly all-encompassing major war script. Instead of focusing on an overarching motive underlying 'ethnic' or 'economy' driven wars, they examine the variation between micro-practices inside conflicts and the conditions in which certain concrete practices occur or do not occur. Thereby emergent dynamics of warfare come to the foreground, such as the relationship between spatial control and types of violence as studied by Kalyvas in his seminal work on the Greek Civil War (2006). Kalyvas and Kocher (2007b), furthermore, pointed to one huge factor emerging out of warfare that normally drives plenty of people into armed groups: protecting your own life by joining a rebellion. During war, it is far from rational to remain outside of a rebellion and hope to receive the profits, simply because it is safer to join. Hence, the classic 'collective action problem', which points to the profits people could have without taking the 'risk' of joining a movement, is turned upside down in a society at war. In war, the puzzle is not why people take part in a rebellion but, rather, why some do not. Or, as Nordstrom remarked, "The least dangerous place to be in contemporary wars is in the military" (1992, p. 271). Therefore, it is not puzzling that people do not simply hope for the benefits that warfare might bring without actually taking part in the risks themselves as so-called 'free-riders' but that they join and risk their lives in order to have at least a better chance of survival. The highest profit people may hope for by joining is keeping their very own life, not certain economic incentives. In the end, it is too complex to calculate costs and benefits in war, even for scholars working on the basis of rational choice approaches, and the sheer fact of a struggle between life and death always intersects with and undermines all mathematical brainteasers.

In her study of the civil war in El Salvador, Elisabeth Wood (2003) also highlighted emergent dynamics resulting from war itself, such as defiance, pleasure in agency (self-empowerment), and the proximity of armed groups, as reasons to join a rebellion. War changes or intersects with motives and delivers a whole new social reality that comes with its own not necessarily rational dynamics. Incentives and resources that may be collected change their value constantly and oftentimes too drastically to allow us to reckon with them as a private or collective gain. Moreover, the actions of individuals and groups also stem from warfare itself and the social conditions it creates. The motives that precede and last during war differ, and motives alone do not necessarily explain violent actions.

A Rationalist Model of Military Organization

In recent research, a shift has taken place towards the study of the internal functioning of armed groups. Most notably, Jeremy Weinstein's study of the impact that different modes of recruitment have on the abuse of civilians turned scientific attention towards organizational aspects of civil wars (cf. Weinstein, 2007; Weinstein & Humphreys, 2005). In the end, however, Weinstein also deduces organizational modes from an economistic and rationalist model of action, and recruitment strategies from resources available during the inception of armed groups. Thereby organizational aspects of rebel groups are directly caused by their initial economic conditions. Weinstein exemplifies the influence that organizational modes of recruitment have on the behavior of armed groups and, at the same time, sketches the limits of organizational control. The tactics, strategies, and behavior of armed groups, especially towards civilians, are shaped by whom and how groups recruit and which rules they set in place to discipline behavior (Weinstein & Humphreys, 2005, p. 8). Factions recruiting combatants by using material benefits, for example, are more likely to exhibit high levels of abuse towards civilians, while armed groups with fewer resources must rely on recruitment via commitment and are more likely to exhibit less abusive behavior. Using high degrees of personal benefits (most notably money) poses an organizational problem, as a KPNLF member put it in a similar tone to that of Weinstein:

> R: It's easy to recruit and then the KP (.) the KP say (1) the KP sometime we use the same method [recruit by ideology like the Khmer Rouge]. (1) We just say that the country was invaded by the Vietnamese so we have to get them out, (.) you know (1). Sometime we use money, (.) sometime we use the rank (2) if we can recruit some people by ideology, (.) they stay with us, (.) you know, (1) they work very honest with us. (1) But if we use money, (.) when we run out of money that is problem ((both laugh)) but the thing that (.) when we don't have money (.) when we don't have rank (.) the people they love each other not very much (.) prefer (.) prefer not to work together. (.) If we have military rank, we have to pay them, (.) you know. (1) That they have ENVY (.) they doesn't want to work because, (.) you know, (.) his rent is lower or (.) his money is less. ((laughs))

I: So, better recruit by ideology?

R: Yeah. (I-KP1)[1]

As this KPNLF leader already suggests, the type of recruitment determines organizational control. If you recruit using money, money becomes one of your main instruments for maintaining control, and you face troubles if the monetary resources decrease. 'Using ideology' to recruit, according to this logic, means that insurgents are more likely to attract recruits who are willing to overcome material hardships, are highly identified with the group, and are therefore readily invest in the well-being of the movement. However, according to Weinstein, it is even the recruitment strategy that is used at the very beginning of a group's inception that determines organizational control and thereby presets the group's general behavior. "Whether a group is filled with activists or opportunists then constrains the choices leaders make as they organize military operations, govern civilian areas, and struggle to retain their members in the course of conflict" (2007, p. 126). Thus, there is a magical force involved in the beginning of a movement, and conditions at the beginning shape the fate of the group.

This is why, in addition to using rationalist and economistic models of explanation, Weinstein delivers a much-too-static model of martial violence in the end, deriving everything from the material resources that were at hand during the inception of a given armed group. This, for example, fails to account for the variance in behavior between groups that had very similar conditions at inception – as shown by Sanín (2008), pointing to the similar (economic) conditions at the inception of two non-state armed groups in the Colombian conflict and their nonetheless systematic differences regarding their social composition and their internal/external behavior. According to Sanín, these differences can only be understood if the group's different sets of organizational devices are taken into consideration. This means that organizational devices and not just initial economic conditions matter. Moreover, Weinstein states the relevance of the influence that "disciplinary structures" (Weinstein & Humphreys, 2005, p. 9) have on the behavior of armed groups, but he deduces that those structures again result from economic conditions during the initial phase of recruitment. Organizational variables ultimately shrink to mere placeholders for initially available resources. Thus, as Yvan Giuchaoua (2009) has shown in his study of the Nigerian Tuareg rebellion *Mouvement des Nigériens pour la justice* (MNJ), the oftentimes erratic trajectories of armed groups and what such organizations are sociologically made of remains unacknowledged by Weinstein's investigation (Giuchaoua, 2009, p. 22). One major problem is that Weinstein thereby renders leaders of armed group powerless in front of their own members. This leads to the very problematic conclusion that the upper echelon of

---

[1]   Interviews that were originally conducted in English are transcribed word-by-word and contain brackets indicating a pause in speech along with numbers or dots to indicate their duration. The coding within the bracket at the end relates to the system of citation from interviews conducted by the author and Koy-Try Teng that can be found in the Bibliography.

the organization is only responsible for the simple fact that they started a rebellion under economically disadvantageous conditions. Moreover, some actions are labeled as abusive and others as non-abusive conduct during war. Beneath his scientific categories, Weinstein introduces an ethical divide, judging between good and bad rebellions.

In a footnote, however, Weinstein and Humphreys briefly state that they do not "record actual violence committed by fighters during the war" but, rather, record "levels of abuse or indiscipline *not* ordered by superiors" (2005, p. 15; emphasis changed). In short, the authors exclude cases of organized violence and include only cases of violence where organizational control broke down and combatants acted on their own. However, this empirical exclusion, in turn, serves as the basis for the theoretical exclusion of the influence an organization (and its disciplinary structures) has on violence later on. Thereby and in an almost deconstructive manner, the excluded becomes included in its own exclusion. Based upon an exclusion of cases of organizational control, the data shows the predominant impact of conditions during recruitment simply due to the fact that the data has been trimmed to fit the theory. Furthermore, while stating in the same footnote that military units that are provided with a "license" to abuse civilians are "also likely to be more abusive overall" (ibid.), the same becomes forgotten during the final discussion where the authors declare that their findings challenge the view that "high levels of abuse and violence are observed where leaders retain tight control over an efficient killing machine that can be directed at will" (ibid., p. 26). Leading to the somehow paradoxical and empty conclusion that in cases of abusive violence which lack organizational control, organizational impacts can be excluded as a causal factor.

Another problem is the questionnaire used by Weinstein in the field, which tries to make ex-combatants talk about civilian abuse within their own units (published online, see Weinstein & Humphreys, 2003). It is rather likely that particularly those who are highly committed to their movement, the 'idealistic activists', do not speak openly about how civilians were manhandled by their own units. Even though Weinstein's questionnaire tries to circumvent the problem at certain points by using indirect questions, it is unlikely that highly identified and committed combatants would damage their former group's reputation, even though they were at least formally no longer members at the time when the interview took place. As a result, one might even expect that the aggregated dataset contains low levels of *reported* abuse within rebel organizations that show high levels of identification and group cohesion. Agents less identified with the group surely talk less reluctantly about the occurrence of 'bad' conduct and indiscipline. A misguiding result from this bias might be that those who answer 'honestly' are the ones who are former members of 'abusive' units simply due to their honesty. This would mean that the propagated empirical link reflects a methodological shortcoming that is difficult to address when interviewing people on civilian abuse.

The current study, however, maintains that disciplinary practices (of which recruitment strategies are a part) are socially rather than economically differentiated. While it is certainly useful to point to the organizationally counterproductive effects of selective economic incentives, it is at least similarly decisive to examine how recruits are socialized and transformed into combatants inside military organizations as well as their exposure to actual combat situations and how they are handled within the organization (cf. Sanín, 2008, p. 6). Thus, modes of (combat) organization take over the initial motives for recruitment. The point is that this organizational handling of combat corresponds with the habitus of agents. Moreover, especially when confronted with actual combat situations, the willingness of combatants to fight and the cost-benefit ratio of behavior become highly questionable. During combat, things are not rational but marked by the fear and incompetence of soldiers, even by those who are highly trained and disciplined. One might say that rationalist models circumvent the central aspect of war: battle and violence. Thereby rationalist action in warfare remains rather compelling but loses its appeal when confronted with the 'messy' reality of battle.

Chapter 2
Habitus Groups and Power

The term *habitus*, as used by the French sociologist Pierre Bourdieu (1930–2002), aims at analyzing embodied practical schemes that are related to the social position of an agent. Individuals acquire basic patterns of behavior in the environment in which they grow up (usually within the realm of the family) and during their lifecourse, in which they are typically part of various institutions and social fields. People incorporate schemes of perception, thought, and action during their socialization, which are applied, repeated, and thereby habitualized under similar but ever-changing conditions. In doing so, schemes of behavior and thought become socially differentiated. Thus, the habitus can be seen as a "structuring structure, which organizes practices and the perception of practices" according to the social position of an agent (Bourdieu, 2010, p. 170).

An individual's social milieu, family of origin, education, upbringing, and participation in fields with different institutions, as well as past choices, form the set of schemes being incorporated in the body of agents. Schemes function as a tendency or disposition to act in certain ways that are 'embedded' in the habitus. The habitus thereby becomes the totality of those generative schemes operating as a set of implicit knowledge governing the agent's everyday practices and judgments. In doing so, they give such practices a coherence and regularity (Bourdieu, 1990, p. 92). Hence, classificatory schemes structure practice and are "able to organize the totality of an agent's thoughts, perceptions, and actions by means of a few generative principles" (ibid., p. 110).

Although they are generated during the lifecourse of individuals and (re)constructed under ever-changing conditions, agents misrecognize the arbitrariness of these classifications. The social world becomes the 'natural world', in which almost everything, especially rules of behavior and thought, is taken for granted – a process Bourdieu calls *doxa* or, with respect to stakes within particular social fields, *illusio* (Bourdieu, 1977, pp. 159–71). In our case, for example, commanders take the 'nature' of rank-and-file soldiers as well as how they should be treated and disciplined for granted. Their attitudes and classifications in everyday judgments come without saying and bring coherence to their thoughts and actions. In the end, this coherence makes certain judgments and actions on the part of the agents likely (but not necessary) because of the tendency of the habitus to reproduce the conditions of its own formation. This universe of symbolic codifications and classifications forms a discourse, as the French philosopher Michel Foucault would call it, a system of representations, which is taken for granted and prescribes what can be thought, said, and done, and how (Foucault, 1970; 1981). Thereby this classificatory discourse with its schemes becomes a 'structuring structure' that governs how topics can be addressed in a meaningful way so that they 'make sense'.

Furthermore, it also tells the agent which actions make sense and how 'things should be done' – making a discourse not just a textual universe but something that intertwines the symbolical representations with practices and institutions. This way, rules are established according to which things are regarded as normal. For example, rules establishing how to 'correctly' punish indiscipline or prepare for combat in order to make a recruit docile. A classificatory discourse on life as a soldier and on military leadership thus frames disciplinary practices and institutions.

To a certain degree, this discourse predetermines the potential course of an agent's action and practices. In everyday attitudes and judgments, this symbolic universe forms an undisputed doxa leading to almost self-evident practices. By connecting the idea of discourse to Bourdieu's set of symbolic codifications, the discourse becomes socially differentiated and dependent on social groups. This makes discourse a part and an instrument of social struggles in society, particularly in certain fields. However, while some discourses are more or less restricted to one group, others can be quite hegemonic within certain societies or encompass certain social spheres. Low-ranking soldiers, as well as some of their commanders with backgrounds in rural farming, share some ideas, such as, for example, the concept that a commander is a model who is bound to soldierly and spiritually guided ethics. Discourse is not just a master cleavage (fighting against Vietnamese) but is socially differentiated into various smaller, overlapping, minor, dominant, and contradictory cleavages, struggles, and classifications.

Field Theory: Access, Valuation, and Struggle

Each social field has its own logic and set of rules. A field reproduces not simply the social structure of a nation state but also, depending on its history, its rules, and thus the field position of its agents can differ sharply. Furthermore, the field of interest for this study sets different rules of valuation due to its unique interlacement between local networks and national divisions, as well as international and global forces along a contested border between Thailand and Cambodia. But first and foremost, it is the condition of a field of organized violence that sets special rules of valuation. War introduces new classifications, valuations, resources, and practices of its own, and it carries over and disrupts older categories. For example, agents can rise in field position depending on the stock of battles they fought as a symbolic resource that is of particular value almost exclusively in the field of resistance.

The resources that agents acquire are valued differently according to the respective field. According to Bourdieu, there are four main types of resources agents can acquire (Rehbein, 2011, p. 278; Bourdieu, 2010). *Economic resources* can be any kind of possession with a monetary or simply an exchange value in the field. In the case of the resistance and due to the breakdown of the economy that included the destruction of money under the Khmer Rouge and an unstable inflation of the new Cambodian currency, the main economic resource was gold. Moreover, goods in the camps such as food cards or fuel for cooking had significant economic value. *Social resources* are relevant connections to people holding influential positions in the field – in our case, mostly siblings or patrons who function as 'sponsors'. *Cultural resources* are all sorts of skills, knowledge, education, and the possession of culturally relevant objects. In the military field, military socializations are highly valued, as is the level of formal education in high school and university. This, however, was very different with the Khmer Rouge, whose anti-intellectualism led to the promotion of poor peasants to high field positions.

The fourth resource is *symbolic*, such as symbolically codified superiority, that is, a superior symbolic value within a field. Each group in the field tries to promote one resource, most likely the one they dispose most of, such as having fought many battles (strongmen) or having a high level of education (intellectuals), in order to legitimize their own power claims within the field. This particular resource in turn also structures an agent's symbolic classifications and practice. This means that symbolic resources correspond with the resource agent's dispose most of, thereby legitimizing their power claim and furthermore structuring the agent's symbolic universe and its practices. Unlike other resources, a symbolic resource is a bit like a meta-resource, in which resources are acknowledged by others in their superior symbolical value and can be used to enforce power claims within a field. In our case, an example of a symbolic resource could be the use of soldierly skills (a cultural resource) to claim one's symbolic superiority as an invulnerable and strong warrior. Within other fields, this claim and symbolic value might not count at all. Thus, symbolic resources reflect the specific structure and rules of valuation within the respective field. Therefore, some symbolic resources are more or less bound to a single field while others remain highly valuable across different fields. For example, during conflicts and due to the history of our field under study, surviving many battles and the genocide as an ordinary soldier became a resource that could be used to facilitate gains in social status among the fighting forces.

Resources are basically acquired in one's family of origin but can change during one's lifecourse and while participating in different social fields. Although acquired skills and resources can be used in other fields as well and may serve as entry criteria, they are mostly not valued and cannot be used in the same way. Thus, the field position is an interplay between the habitus of an agent with his set of resources and the rules of valuation that define the entry point and possible course and pace of mobility in the field. For example, the recruit's initial military rank changes according to the degree of social resources or contacts he has to the top leadership. Rank promotion and mobility in the field, furthermore, are a result of the acquired resources and specific rules set by dominant habitus groups. Rank-and-file soldiers of the non-Communist resistance, for example, were classified as incapable of personal improvement, which is why they received few promotions in rank other than when they were given certain 'positions', such as leadership of a small group (*puok*).

Field rules, however, are not set in stone. They are highly contested by all agents in the field. Very different agents can suddenly find themselves in a position where they have to compete with others who 'normally' are not in a similar social position. This was the case in the military resistance in Cambodia, for example, where intellectuals and politicians from top government positions in the Khmer Republic suddenly found themselves having to compete for leadership positions with comparatively poorly educated military leaders. Yet there is just one overriding principle governing the field, which is not contested and needs to be tacitly acknowledged by the agents. Bourdieu calls this principle the *nomos* of a field and its tacit acknowledgement by the agents within the *illusio*. Seen by an observer from the outside, this *illusio* may look incomprehensible or even strange. In our case, this would be the tacit knowledge that it is worth fighting the Vietnamese, and in many cases even with great personal investment and high stakes, in which people die for the cause. Participation within the field is bound to the acknowledgment of this *nomos*, which is why some of the respondents who did not join to fight the Vietnamese oftentimes only ironically referred to their 'high motivation' to 'save the country from the Vietnamese'. This, however, only occurred among some low-ranking recruits, not among the political and military leadership who incorporated an eloquent version of the political necessity to fight, despite the fact that they did not do the actual fighting. Commitment to or acknowledgment of the value of the activities and resources circulating within the field (*illusio*) serve as prerequisites for an agent's participation in the field.

A field also has no fixed boundaries. While Bourdieu would say that a field ends where its effects cease, agents within the field are also in a constant struggle to define what and who is part of the given field. A practical example can also be found in the role of the Khmer Rouge within the field of resistance. Because the Khmer Rouge were former enemies, with whom many still actually fought, some soldiers did not acknowledge them as members of the field but saw them as potential enemies. The habitus is always relational, be it a single or a group habitus. Agents and groups occupying positions in the field always try to make a (more or less symbolic) distinction between themselves and other groups in order to legitimize, enhance, and secure their own position in society or in a particular field. In doing so, an individual's classifications of others, for example of rank-and-file soldiers, thereby also indicate a relational positioning and symbolic struggle by the speaker himself (the commander). As in physical fields, one's own position and one's relation to others defines each element. Furthermore, the agent's positions in the field are accompanied by different degrees of influence (or 'weight') regarding how the rules are set up, altered, and to be followed. Hence, power is as relational as is any practice in the field and cannot be analyzed outside of the field's contested relations of force.

Microphysics and Power Meshes inside the Field

In his analysis of the 'microphysics' of power, Michel Foucault highlights the need to understand power as relational rather than as a substance or 'property' accumulated by subjects in order to 'make use of it' (Foucault, 1991, pp. 26f). This means that power should be conceived as a field of forces, similar to Pierre Bourdieu's relational thinking, in which constant struggles take place and where everything can only be seen as a force in action while being exercised. Power is nothing mysterious but is a social relation within a field of contestation. It is "not a substance. Neither is it a mysterious property whose origin must be delved into. Power is only a certain type of relation between individuals" (Foucault, 1981, p. 253).

Depending on the individual's field position, power practices are more or less hegemonic and able to define modes of access, like criteria for military recruitment, and to form institutions and binding rules for others in the field. Contestation was common among leading habitus groups of the military resistance but also between leadership and subaltern rank-and-file soldiers. That means that power is never exhaustive and leaves room for seemingly powerless subordinates to develop their own strategies for securing power: "The characteristic feature of power is that some men can more or less determine other men's conduct – but never exhaustively" (ibid.). Although leading groups can define hegemonic power practices and institutions, lower and subaltern agents still take part either in their own subjugation by making themselves docile or they undermine discipline using disciplinary micro-practices. A simple example of undermining discipline would be refusing to obey orders, which can be done more or less openly, especially during a battle situation. However, at the same time, soldiers may take part in their own subjugation by, for example, engaging in practices to decrease their own fear or combat reluctance of any sort. Thereby the rules become manifested and extended by those who are dominated. Such patterns of power practices are also framed by the incorporated discourse of the rank-and-file soldier, which defines rules pertaining to when and how power is undermined or manifested by micro-practices.

So far, Michel Foucault's analysis of the microphysics of power entails a framework of relational thinking that is similar to Bourdieu's field theory. Individuals always think and act from a certain position in and against the field of forces, which is why a discourse as well as practices always carry the imprint of a relational positioning and mirror the social position of the respective individual. However, Foucault's analysis of different power types or rationales is scattered over several of his works over the years and is particularly important for the study of the field of resistance here. Foucault roughly differentiated between five power rationales that are relevant for the purpose of this study: sovereign, disciplinary, and pastoral power as well as security and self-techniques. Sovereign power refers to the reconstitution of personal loyalty to sovereignty via symbolic caretaking, force, and blood, in which the sovereign individual or entity is the only source of justice and goodwill. He is the one who decides between a binary choice of good and evil, correct and wrong, as he disposes of the absolute power to decide accordingly. The sovereign, furthermore, is the moral center of the fiefdom and diffuses good conduct to his allegiance (Foucault, 1991; 2009, p. 28). Subordinates are divided into those who are unconditionally loyal and those who are not, with rewards and punishments used to reconstitute sovereignty when it is threatened by disobedience.

Disciplinary power operates as a corrective process of normalization using a model to guide its operations:

> Disciplinary normalization consists of first of all in positing a model, an optimal model that is constructed on terms of a certain result, and the operation of

> disciplinary normalization consists on trying to get people, movements, and actions to conform to this norm, and the abnormal that which is incapable of conforming to the norm. (Foucault, 2009, p. 85)

In disciplinary power, the subject is under constant surveillance and corrective pressure to be modified according to a preset norm. Punishment thus is not binary as in sovereign power but becomes essentially corrective on a continuum: "To punish is to exercise" (Foucault, 1991, p. 180). Thereby the subaltern can be made or transformed into a norm – or fail and classified as abnormal. Everything is broken down into small components in order to establish complete visibility of the individual and to incorporate the norm into the habits and gestures of the subject as disciplinary techniques for soldiers in late eighteenth-century Europe exemplify:

> [T]he soldier has become something that can be made; out of formless clay, and inapt body, the machine required can be constructed; posture is gradually corrected; a calculated constraint runs slowly through each part of the body, mastering it, making it pliable, ready at all times, turning silently into the automatism of habit; in short, one has 'got rid of the peasant' and given him 'the air of a soldier'. (Foucault, 1991, p. 135)

While disciplinary techniques transform subjects by way of constant surveillance combined with corrective practice, pastoral power uses the technique of confession to modify subalterns according to the will of the ruler. Confession aims at the renunciation of the self in favor of the leader's will (Foucault, 1981). Similar to disciplinary power, pastoral confession aims at a transformation and modeling of the individual subaltern, in which he must acknowledge his wrongdoings and shortcomings in order to change his behavior and thoughts in a next step. Confession thereby practices a constant mortification of the 'deluded' self.

Security aims not at the subjugation of the individual but at the governance of collectivities or the 'management of flows'. That is, it manages multiplicities as series of unchangeable given elements, it "works on the given", thereby maximizing the positive effects and minimizing "the risky and inconvenient" effects (Foucault, 2009, p. 34). Rule becomes governance that attempts to regulate multiplicities of any sort (like goods, people, or information) and control their mode of circulation and its collective states. Measures are tested and monitored by the effects they yield on the respective collective group. Punishments, for example, do not aim at revenge or the reconstitution of sovereignty but weigh the effects that crime and punishments have on certain collectivities (in our case, on the organizational coherence or the civilians' perception of the movement).

With self-techniques, Foucault points towards processes in which subjects transform themselves according to a power force or power is transformed using minor strategies that the subjects apply in order to adjust it to their own will – more or less hidden strategies of resistance. These techniques can be used by anyone but are typically found among subalterns who face disproportional power forces. By either adjusting the self to the rules or the rules to the self, previously intolerable conditions can be made more bearable for the subaltern. This is the case, for example, with soldiers using religious protection or rituals to decrease their fear during combat. In doing so, a situation in which they feel forced to do something becomes legitimate, and the subjugated adapt to fulfill their symbolic position with all its 'necessities', a process Bourdieu would call 'symbolic violence' (Bourdieu & Wacquant, 1992, p. 143). The soldier, for example, reasons that it is 'normal to be a soldier for people like him' and that he 'has no choice but to deal with his fears and comply'. Soldiers who, in contrast, change the conditions instead of themselves try to avoid direct combat or shooting by exhibiting minor, almost undetectable forms of indiscipline. While Foucault aims to analyze the "art of existence" that transforms one's existence into an oeuvre, for the purposes of the current study, I will limit the use of the concept of self-techniques simply to "reflective and voluntary practices by which men not only set themselves rules of conduct, but also seek to transform themselves" or the situation they are facing (Foucault, 1992, pp. 10f). Self-transformative techniques can thus be seen as a form of symbolic violence, in which arbitrary and socially constructed inequality is reinforced by the dominated (which I would call *self-rectification*), or as an attempt to resist the demands of the power holders (or: *strategies of resistance* or *lines of flight*[1]).

Thus, it is not solely the leadership that shapes power practices; rather, each agent in the field takes part in shaping discipline according to classificatory schemes embedded in their habitus. Power practices in habitus groups become aligned along one of these power rationales, thereby forming a combination of knowledge (classificatory discourse) and power (rationale), which Foucault calls *meshes of power* or a *dispositif* (Foucault, 2007). A dispositif is the totality of discourses, practices, institutions, rules, knowledge, and power techniques that, in our case, characterize the thoughts and actions of different habitus groups. Each group forms a mesh of practices and discourses that are embedded in the relations within the field and its accumulated history. Practices are formed by the respective discourse and the rationale dominating the habitus group's power field. However, with traditional patterns of rule, there is an additional third force that shapes power practices inside armed groups. These traditional patterns of rule can be called 'socio-cultures' and relate to the habitus as a "tradition inscribed in a person" (Rehbein, 2007, p. 279). Hence, older layers of rule and patterns of domination connected to them persist over time and are constantly reinforced.

With regard to the field of military resistance, there are two interrelated patterns that shape power practices among and between many groups: patronage and subsistence ethics. In patron-client relationships, the clients seek the protection of powerful patrons who are able to guarantee their basic subsistence and security. Instead of striving for profit, the clients' interest in subsistence ethics is focused on securing his or her livelihood in an exchange relation with a patron by reinforcing family ties or by mutually aiding or assisting each other (reciprocity) (Scott, 1976). The patron in return strives to enlarge his base of support by securing as many dyadically reinforced loyalties as possible. Because he guarantees

---

[1]   'Lines of flight' refers to Gilles Deleuze and Félix Guattari and their critique on an implicit fatalism within Michel Foucault's concept of power (2004).

subsistence against all odds and insecurities, especially under conditions of war, the clients give him their political (and in our case: military) support. Patronage is a traditional pattern of rule that remains widespread in Southeast Asia today. However, it became especially important in the field of military resistance due to the increased need for protection and basic goods to which many people did not have access without a military patron who demanded their service as soldiers in return. While some of the characteristics of patronage are common to some habitus groups, patronage practices are also socially differentiated. The practice itself becomes multi-centric and changes its meaning and contents according to the respective group.

This study assumes that the different habitus groups share a distinctive discourse, which can be analyzed using a qualitative approach developed by the study group of Vester, Vester-Lange, and Bremer (e.g., Bremer, 2004) that was adapted by Rehbein (2008) for the study of global inequality and for the purpose of the current study on a field at war. Hence, in contrast to Foucault, it assumes that the discourse is socially differentiated and allows for a hermeneutical analysis of the habitus and its formation. To do this, the interviews focus on the respondent's social background, the family of origin, the lifecourse and acquired resources, and classificatory schemes on being a soldier and leadership, as well as power practices. A methodology that captures social forces and agents who defy the limits of a nation state and its occupational divides fits the demands within a study of a civil war, in which agents and institutions defy the nation state as well with four decisive types of resources, which are difficult to quantify and pin down in their value. By keeping the questions as open as possible, respondents are able to decide which lifetime sequences and frameworks are most important to their own biography and how to frame their classificatory schemes. Instead of elaborating on a preset social position within the interview, the social position and habitus becomes part of the interview analysis and is set by how the respondent marks differences in his own words. Habitus groups are constructed according to the entire set of factors, such as shared social backgrounds, shared lifecourses, shared discourses, and shared power practices. Later on, they are additionally grouped according to the military rank within the resistance and its organizational hierarchy without claiming a quantifiable distance between positions as Bourdieu did. Chapter 7 will further elaborate upon theoretical deviations from Pierre Bourdieu's as well as Michel Foucault's framework that are needed for a sociological perspective on civil wars and conflict and that emanate from the following analysis of habitus groups within the Cambodian insurgency in particular.

Chapter 3
Field History and Structure

The Thai border has a long history of serving as a refuge for Cambodian resistance groups. Anti-colonial fighters of the *Khmer Issarak* and *Sereikar* ('Free Khmer') operated in the remote areas of the northwestern border regions during the 1940s and 1950s to fight for independence from the French. Some Communists, supported by the Indochinese Communist Party with close connections to the Sereikar, were active there as well. Again and again, nationalist and Communist groups retreated to this 'no man's land' when they were chased by state militaries, the police, or other armed groups.

When the Communist Khmer Rouge took power in 1975, officials, policemen, and soldiers from the Khmer Republic either immediately went abroad to the US or France, took refuge at the border, or tried to hide among the populace for several months if not years. However, most members of the *Lon Nol regime*, as the Republic was termed with reference to the prime minister and president, were singled out for execution at the very beginning. Those who managed to hide their identity became spectators and victims of the Communists' attempt to rebuild the society into an agrarian-industrial and egalitarian utopia with forced evacuations of city-dwellers, forced labor on the rice fields or in irrigation projects resulting in famines and starvation, and widespread purges of 'enemies of the people', mainly organized by way of an extensive detention center system, leaving an estimated number of 1,7 million people dead (cf. Kiernan, 2002; Bultmann, 2012). For the party's leadership surrounding Saloth Sar, better known by his nom du guerre 'Pol Pot', the new state of *Democratic Kampuchea* was to be cleansed of capitalist exploitation, colonialism, and feudalism, fighting against 'Khmer bodies with Vietnamese minds' who were resisting the new collective order. An extensive system of detention centers presided over re-educating or 'smashing' strings of 'traitors'.

Weakened by starvation and mostly unarmed, many fought their way to the border region in order to meet with one of the many but small and factionalized groups resisting the Khmer Rouge. Probably the strongest force between 1975 and 1979 was the former 13th Brigade of the Republican Army (FANK) under the command of Prince Chantaraingsey and Kong Hok Tranh. The brigade remained active until about 1977, with around 2,000 men (cf. Corfield, 1991). Other resistance groups at that time included Khleang Moeung, Cobra, Khmer Liberation Movement, Khmer Islam, Sereikar Oddar Tus, and a group surrounding the former Republican prime minister In Tam. But all became more and more factionalized over the years while fighting without major support and being chased not just by the Khmer Rouge but also by Thai police and military During the early years, many groups were either on the brink of extinction, became small-scale smugglers with Thai patrons, or degenerated to mere intelligence units for the Thais on Khmer Rouge border activities. Some small bands were still active with tit-for-tat raids along the border (Conboy, 2013, pp. 115–17). Others worked as protection units for Thai peasants against Red Thai guerrilla along the border (BC-F5). Some even searched for resistance groups they could join but failed to find them, living alone in the woods for a couple of months and ambushing small Khmer Rouge patrols for food and weapons. However, these small groups became the backbone of some regiments and brigades later on. Thereby many rather low-ranking soldiers climbed in rank because they were the first ones along the border and the first 'force' on which people could rely.

At the end of 1978, border clashes with Vietnam increased and Vietnam and defected Khmer Rouge soldiers prepared for an attack on Cambodia. On December 25, 1978, about 100,000 Vietnamese troops and 20,000 soldiers from the *Kampuchean National United Front for National Salvation* (KNUFNS), consisting predominantly of defected Khmer Rouge soldiers headed by Heng Samrin, launched an offensive designed to topple the regime in Phnom Penh. Due to the low morale within the Khmer Rouge military apparatus and the widespread starvation of the populace, the Vietnamese invasion was relatively quick and successful after only a few weeks, with Phnom Penh being captured on January 7, 1979. While most parts of the country fell to the Vietnamese within a short time, the core of the Khmer Rouge political and military apparatus managed to flee to the border regions of Oddar Meanchey. Although many died or defected, the leadership was nevertheless able to take hundreds of thousands of civilians and a considerable number of cadres with them, mostly under threat of being killed by the Vietnamese if they stayed behind: "They always said: 'If you go back [to the interior], you will be killed. They are only looking for Khmer Rouge soldiers'. Hearing so, we did not dare to go" (BAC-KR1).

While some went into hiding in the interior, the bulk of the cadres were reorganizing along the border in preset camps with some stored food and weapons or mixed among ordinary refugees seeking protection and food from the Thai. Some members of the Central Committee were flown to Bangkok and headed further to China to meet their patrons. Although they tried to destroy the remnants of the Khmer Rouge, the Vietnamese failed because they could not identify all cadres among the refugees in the midst of the chaos in the aftermath of the invasion and most notably because of the military support by the Thai and Chinese. While the Thai feared that the fighting would spill over into their own territory, therefore pushing back Vietnamese units, the Chinese 'taught the Vietnamese a lesson' along the northern Vietnamese-Chinese border in order to lift the pressure from their Cambodian comrades (cf. Chanda, 1986; Evans & Rowley, 1990). Back in Phnom Penh, the Vietnamese installed the *People's Republic of Kampuchea* (PRK), which was staffed mainly with former Khmer Rouge cadres who had defected across the border before internal purges staged by their Khmer Rouge comrades reached them in late 1978.

Being backed by the Soviet Union meant that not only China but also the US and other Western states refrained from offering support or any sort of legal recognition for the Vietnamese 'puppet regime'. Instead, the Khmer Rouge and Sereikar along the border functioned as an attractive military and political force against the 'hegemonistic expansionism' of the Soviets in Indochina and would at least keep it at bay without further involvement of US troops (an option that was not highly favored by the Americans after the Vietnam War). Thereby Cambodia became intertwined in Cold War logic until 1989, at which point the US did its best to support any force resisting Soviet-backed states. Although the horrors under the Khmer Rouge regime became widely known (especially due to the Oscar-winning movie 'The Killing Fields' in 1985), the US stuck to their policy of Cold War *realpolitik*. However, the civil war after the genocide lasted a decade longer than the Cold War until the final surrender of General Ta Mok in 1999, already indicating that it was much more than merely a war by proxy.

Early Years in the Field: Refugee Crisis and Aid

In mid-1979, more and more refugees came to the areas with Khmer Rouge soldiers and non-Communist resistance groups scattered along the border. At the beginning, the Thai were unable to cope with these masses of people and forced more than 40,000 refugees back over the Dangrek mountains to the interior of Cambodia into a heavily mined and malaria-ridden territory, even killing those who refused to leave and thereby causing thousands of deaths by starvation, sickness, and land mines (Lawyers Committee, 1987, p. 26; Robinson, 2000, p. 24; Deth, 2009). However, after realizing that the harvesting season would not provide enough yields because rice was not planted during the chaotic aftermath of the genocide, the number of refugees again increased considerably. Forced repatriation remained the main reaction by the Thai government when faced with the first flows of refugees. It took a couple of months until aid for thousands of starving and desperate people who were knocking on Thailand's door slowly became institutionalized and met the basic needs along the border. Meanwhile, chaos mounted in the Cambodian interior, where everyone was on the move after the genocide, desperately looking for food and relatives. Only after the visit of the Thai premier to the border did the government allow distribution of first aid to the refugees and establishment of the first temporary holding centers for about 40,000 people.

Still on the Cambodian side of the Thai border, the resistance leaders of the Khmer Rouge and the Sereikar, along with the Thai state, asked for aid from international donors, relief organizations, and the UN refugee agency *UN High Commissioner for Refugees* (UNHCR). By the end of 1979, all appeals – by aid agencies who engaged in uncoordinated efforts to help the refugees, by the Thai, and by rebel groups – to international donors finally succeeded, and the Cambodian refugee problem became widely known to the world through media reports on all channels showing starving and dying children. At first, however, the door to Thailand for those seeking asylum abroad was only briefly open. At the end of 1979, UNHCR set up a camp in Thailand, called Khao-I-Dang, and some other centers in which Cambodians could seek resettlement to 'third countries' such as the US or France. Already by early 1980, however, the center was officially closed to new arrivals, and the bulk of the refugees in various settlements along the border were declared 'displaced persons' or 'illegal immigrants' and only insufficiently assisted by UNICEF and Red Cross (Robinson, 2000, p. 28). Exact numbers of arrivals are impossible to assess and fluctuated heavily, but a Finnish Inquiry Commission estimated that there were approximately 850,000 refugees at the border between 1979 and 1980 (Rogge, 1990, p. 32). Because some people sought food but then returned to their home villages, it was even more complicated to assess reliable numbers.

Finally, as late as March 1980, the first land bridges were established to distribute rice and goods to the refugees residing behind the border line. Land bridges were spots along the border where trucks with tons of rice met with streams of Cambodians from the interior on oxcarts and bikes who were waiting for the distribution and then went back to somewhere behind the border – to refugee settlements or simply home. With considerable support from the UN, as well as increasing aid from all over, Thailand agreed to organize coordinated support for the refugees from that point on instead of sending them back into malaria-ridden and heavily mined areas behind their own border. The Thai government started to grant refugees a shelter, aid agencies an operational base for supplying the land bridges and camps, and sanctuary for rebel groups as well. The Thai changed their policy towards the refugees and gave the Khmer Rouge and the Sereikar protection, mainly due to the presence of the Vietnamese troops in Cambodia, who showed up along the border as well, searching for the remnants of the Khmer Rouge. The government considered them a direct threat and feared a spillover of hostilities to Thai territory or even a further 'expansionistic' drive by the Soviet-backed Vietnamese. Hence, together with the UN, China, the US, ASEAN, and others, the Thai government organized support for the rebels from that point on by providing shelter and logistical assistance along the border in order to keep the Vietnamese at bay. Therefore, they allowed relief workers and agencies to operate in that area. Moreover, for more than a decade, the Khmer Rouge and their allies were granted the Cambodian seat at the UN Security Council. Weapons, money, food, and all kinds of material were trucked to the border through various checkpoints that collected some extra money for the passage-way. The refugees inside increasingly militarized camps alongside thousands of aid workers became a useful buffer to fight the Vietnamese.

For the aid agencies, however, this political constellation created a difficult moral dilemma: The distribution of emergency aid would save many lives but also directly benefit and feed the guerrilla factions, leading to attempts by UNHCR and other agencies to limit their services (Mason & Brown, 1983; D34027[1]). But how could the UNHCR dare to openly oppose feeding Khmer Rouge and Sereikar soldiers when the UN and most of the Western states officially recognized them as the government of 'Kampuchea'? Officially, many agencies started to ignore their

---

[1]   These numbers refer to the catalogue of the *Documentation Center of Cambodia* (DC-Cam) in the reference list.

presence. Incoming flows of food meant sudden and almost unconditional power for the Sereikar leaders, who established control over rice distributions and were thus in a position to decide over life and death. In early 1982, the *UN Border Relief Operation* (UNBRO) was created in order to organize relief operations within the scattered refugee settlements and into the interior via land bridges. UNBRO also provided a centralized operational umbrella for all of the different relief efforts and subsequently established refugee camps under the nominal but formal control of the UN. Another reason for establishing UNBRO was simply that the resistance had been recognized as the legitimate government of Cambodia, which is why the population in the camps was then formally regarded as 'citizens' and no longer 'refugees'. As a *refugee* commission, UNHCR was not mandated to operate within this framework. UNHCR only presided over the camp Khao-I-Dang where these 'citizens' could apply for resettlement. Since the UN officially acknowledged the resistance as the legitimate government of Cambodia, people working for the UNBRO apparatus were ordered to remain silent on the military activities of the resistance.

Almost every one of the twenty-one camps[2] along the border was under the de facto control of a non-Communist armed group or the remnants of the Khmer Rouge (see map in Appendix I). Yet most of the non-Communist groups remained rather autonomous for a couple of years while they were officially on the Cambodian side of the border, presiding over the flows of goods and people and making use of the geopolitical position they inherited and defended during their struggle against the Khmer Rouge: "The front was not established yet. It was still rather like a tribe" (RC-KP2). The influx of people and goods from international donors and from smugglers turned their misery into luck and made them strongmen in control of black markets and constantly growing refugee encampments, seen as little 'tribes' or 'fiefdoms'. Former soldiers with only few weapons could make use of this small window of opportunity and gain control over the incoming flows. Recruitment became rather easy, with hungry men arriving at the camps and some force being added to make them stay. Step by step, however, three remnants of former Cambodian governments that had toppled each other over the years were reconstituted along the border, creating a resistance front whose loyalties reached back to the political elite of the *Sangkum Reastr Niyum* under former King Norodom Sihanouk (1955–1970).

(Re-)Birth of Political Parties I: The Khmer Rouge after the Khmer Rouge

Starting towards the end of 1979 and in full operation in early 1980, the remnants of the Khmer Rouge began their 'second life', made possible by the shelter and food provided at the border (cf. Rowley, 2007). They were fed by international

---

[2] A major Vietnamese attack during the dry season in late 1984 and early 1985 forced the population over to the Thai side and reduced the number of camps to eleven, with separate camps for civilians and the military.

aid and survived near extinction while seeking refuge at the Thai border (cf. Shawcross, 1984, pp. 328–61). Being on a drip of international support and with news spreading all over the world about their policies resulting in organized mass murder and widespread starvation leading to millions of deaths during their years in power, the leadership hurried to give the Communist party a new public face. While the leadership was seeking international support around the world, encampments and 'liberated zones' were established along the border, many of which had already been prepared in anticipation of the Vietnamese invasion in order to receive the retreating troops.

After reorganizing (or more precisely, revitalizing) the military and party structure in order to fight a guerrilla war and securing broad international support, the 'new' Khmer Rouge announced that from then on they were going to strive for liberal capitalism and a parliamentary democracy (cf. Peschoux, 1992). Officially and on the surface, the party adapted to the new situation. Instead of *Communist Party of Kampuchea* (CPK), the party called itself *Party of Democratic Kampuchea* (PDK) from December 1981 on, with the *National Army of Democratic Kampuchea* (NADK) as its military wing. In 1985 Pol Pot even resigned to an 'advisory role' due to his 'age', handing over party leadership to Khieu Samphan, with Son Sen as NADK commander-in-chief. Furthermore, the black uniforms were changed to olive-green. And even the Communist terms had to be removed from the party's language, as an internal document of the Khmer Rouge highlights:

> [A]ny word that is too blatant and of the old Communist regime and gives a connotation raising suspicion that we do not change must not be used. We use any word suitable to the situation that the Communist Party of Kampuchea has been dissolved. For example, the word 'Mitr' for comrade is sometimes used in our radio broadcasts. (KR file, 1982, p. 149)

All Khmer Rouge policies became double-faced. For example, as a 'survival guide' (BAC-KR1), a "no more killings order" was implemented that resulted in considerably less organized murder. However, there was still evidence of internal purges, such as a mass grave near the Khmer Rouge northern stronghold Anlong Veng that was found by the Documentation Center of Cambodia (DC-Cam) and held 3,000 victims who were killed between 1993 and 1997 (Rowley, 2007). Furthermore, most observers question the degree to which there was a real reshuffling of offices in the party or whether in reality Pol Pot still held all power (cf. Peschoux, 1992, p. 35). But the reorganization was not merely a façade. It also reflected a change from politics to armed struggle, in which the party's ruling body, the Standing Committee, was disbanded for a 'Military Directorate' (cf. Short, 2004, pp. 417–19).

The two-faced policies of the Khmer Rouge were also geographically reflected in different zones and camps. While the area at the western border with the camp Site 8 and later on the 'liberated zone' of Pailin was relatively open to the public as well as rather 'capitalist' and 'liberal' in attitude, making a lot of money with gem mining, the northern regions surrounding Anlong Veng and Preah Vihear were strictly closed to outsiders and followed old policies resembling popular control during the years of Khmer Rouge reign. Especially Site 8 was designed as a model camp to satisfy the public and the international observers, presenting a new 'democratic' and 'capitalist' face. While Pailin made a comparably rich impression, Anlong Veng seemed to be relatively impoverished.

At least the Khmer Rouge managed to regain some strength. After being toppled by the Vietnamese, about 40,000 combatants remained at the border, and the number lingered at a constant level between 30,000 and 40,000 during the 1980s and 1990s. Some former companies and platoons from the military even remained active or in hiding in the interior for a couple of years until they managed to reach the border as well. Other areas, for example Veal Veng in the southwest, continued to be strongholds for the Khmer Rouge over the years, although far away from the border camps. The command remained largely intact, and old structures were revived quickly.

While some policies were implemented to adapt to the new situation and to win the 'hearts and minds of the people', most camps and zones under Khmer Rouge control resembled old policies of isolation from the outside world, restriction of movement, a strict command economy using food to subjugate the subalterns, denial of individuality in favor of complete subjugation under the 'collective', segmentation of units to control the flow of information, harsh punishments for minor mistakes, and, last but not least, murder of undisciplined 'dissidents' or 'hidden Vietnamese agents'. However, in order to win the hearts and minds of the people, more emphasis was placed on nationalism and liberation rather than Communism, although most of them were certainly won by instilling fear. But here as well there were deep differences between different areas of Khmer Rouge control and certain commanders. The Khmer Rouge was certainly not just a uniform ideological and organizational bloc. There were different factions and different interpretations of norms, practices, and political directions, especially among the members of the leadership.

Therefore, there is discussion surrounding the question whether Pol Pot was in control of everything and to what extent the movement was actually unified. Perhaps the striking organizational difference between the regions of Pailin and Anlong Veng, for example, could be taken as a sign of a cleavage, but it could also have simply been a policy used to distract national and international observers. It seems like the question of central control is highly ideological among observers. While some saw a 'shadow state' at work, in which Pol Pot was secretly in charge of everything with "available evidence" that he never effectively stepped down (Peschoux, 1992, p. 30), others, like the CIA, interpreted his abdication in 1981 as a sign of conflict (CIA, 1987, p. 5). Journalists even reported scattered occurrences of clashes between Pailin and Anlong Veng (Bekaert, 1997, p. 219).

(Re-)Birth of Political Parties II: The KPNLF

The *Khmer People's National Liberation Front* (KPNLF) always emphasized its set-up as a 'front' uniting various factions fighting along the Thai border, having only resistance against the Vietnamese (or previously, the Khmer Rouge) in common. This military and political 'strategy' was termed 'ocean policy', in which all different flows of interests and groups were gathered under one unifying organization (cf. Kong, 2009). For the political leadership, lack of unity was not a weakness but an explicit strength of the KPNLF, mirroring the overarching goal of uniting Khmer from all geographical, social, and political strata. Although the very top of the political leadership was primarily a successor of the former Republican elite reaching back to the Democratic Party during the 1940s and 1950s as well as the Khmer Republic from 1970 to 1975, many other factions found shelter in the KPNLF, making the political direction much less clear and providing a basis for a long leadership quarrel during the mid-1980s. This united front strategy also led to a lot of contestation in setting up binding rules for the field.

Former Brigadier General Dien Del went to the border on February 1, 1979, in order to persuade all of the rather small Sereikar groups to join the *Khmer People's National Liberation Armed Forces* (KPNLAF) under the political leadership of Son Sann, a former short-term prime minister of the Khmer Republic. However, not everyone at the border was in favor of greater unity. Only after Dien Del organized weapons and rice from the Thai did he manage to form some unity among the Sereikar leaders, at least for the time being. Most Sereikar joined primarily because of the chances they saw for international support (money and weapons, in particular) channeled through the KPNLF but remained rather independent in their operations throughout the period of resistance (cf. Corfield, 1991). The downside of the 'ocean policy' was a substantial lack of inner control over commanders who were frequently engaging in black-marketeering instead of combat. Especially one group resisted being incorporated into the front: the *Mouvement de Liberation National du Kampuchea* (MOULINAKA) under the command of the former Naval Captain Kong Sileah. After his death in 1980 (apparently due to malaria), Nhem Sophon took over the movement, who then preferred the royalist resistance over the KPNLF. Depending on how lucrative the geographical location of an independent Sereikar leader was in terms of goods, people, and weapons, he could survive for a while without the benefits associated with Son Sann's KPNLF. Only after some killings and battles took place between different commanders did each commander find and opt for a protective umbrella for his respective operations.

(Re-)Birth of Political Parties III: Funcinpec

A major factor in the effort to unite these Sereikar was the formation of the *Armée nationale sihanoukiste* (ANS) and its political wing, the *Front Uni National pour un Cambodge, Indépendant, Neutre, Pacifique, et Coopératif* (Funcinpec) in February 1981, loyal to former King Norodom Sihanouk. Moulinaka, the largest still independent force, became the major military faction within the ANS after Kong Sileah and Nhem Sophon resisted integration into the KPNLF. Other smaller groups even defected from the KPNLAF to ANS after formation. While Moulinaka was certainly the biggest force, others included *Khleang Moeung, Sereikar Oddar*

Tus, Khmer Angkor, Praloeng Khmer, and *Damrai Saw Phluk Khiew.* During the first few years, the military command was given to In Tam, who was one of the leaders responsible for the coup deposing Sihanouk in 1970. After only a few years, Sihanouk's son Norodom Ranariddh replaced him as ANS commander-in-chief.

The ANS was clearly the smallest force of all three factions, and with Site B was in control of only one major camp. However, the strength of the ANS rose constantly, especially after the major dry-season attacks of the Vietnamese in 1984–1985 – in stark contrast to the development of the KPNLAF. The ANS profited from a rather unified command and, at least for international observers, rather surprising strength during a battle against a Vietnamese attack on their camp at Tatum (although they were ultimately defeated in the end). As a whole, the party leadership did not really aim for a military defeat of the Vietnamese but was simply in need of a military force that could negotiate with international donors, other parties, and the Phnom Penh government. It was the political strength and appeal of Sihanouk himself that was the core of Funcinpec's strategy – an appeal that was especially important in the international arena of politics that was being addressed by all three factions as they sought to form a joint political body with the *Coalition Government of Democratic Kampuchea* in 1982.

The Coalition Government of Democratic Kampuchea (CGDK)

Under pressure from major states to gain military and political credibility, all three archenemies agreed to form a tripartite government with Sihanouk as President, Khieu Samphan his Vice-President, and Son Sann the Prime Minister. This move secured increasing flows of support from several states such as China, the US, ASEAN, and many more, as well as a seat at the UN until the early 1990s. A coalition, however, more or less existed only on paper and suffered many incidents of military infightings, with the Khmer Rouge constantly attacking their non-Communist partners on the battlefield. As an observer stated, there were four things wrong with the CGDK: "It was not a coalition, it was not a government, it was not democratic, and it was not in Kampuchea" (Robinson, 2000, p. 27). Even worse, all coalition members were officials from three successor regimes that had toppled one another and had plunged Cambodia into civil war and genocide.

Over the years, many political leaders of the CGDK took rather erratic positions on the question of whether or not to join forces. Most notably, Sihanouk's position was hard to follow. After being imprisoned by the Khmer Rouge during the genocide and having lost many of his children and grandchildren to the Communists purges, he attempted to flee from their influence and sought political asylum in many different countries. During this time he also stated the following regarding possible cooperation with the KPNLF: "They will not fight the Pol Potians – they are paid by China! And I cannot pay them. I have no money to pay them" (Schier, 1985, p. 47).

In the end, China was, in fact, the only state that offered him political asylum and eventually pressured him to join forces with the Khmer Rouge and KPNLF. Thereby Sihanouk reiterated a cooperation that he himself had started in 1970 with the Khmer Rouge to topple the Lon Nol regime and that had ended in years of political turmoil and a genocidal disaster that left millions dead. His reaction was that of a victim and patriot with few choices available to him other than to follow historical fate: "I am a lamb. Son Sann is a lamb. We have to choose between being eaten by Khmer or being eaten by Vietnamese. As Khmer, we prefer to be eaten by Khmer, because we are nationalists" (Evans & Rowley, 1990, p. 209). Over the years, he constantly threatened to step down due to repeated ambushes by the Khmer Rouge on his troops.

Not only the former king but also his non-Communist coalition partner, the KPNLF, had a hard time explaining why they were joining the genocidal Khmer Rouge whom they had resisted for so long and who had brought so much suffering to their people. Son Sann drew a very figurative picture while he was touring the refugee camps in order to make sense to this decision: "If we drive and have to turn right, we cannot turn sharply, we have to move left first and turn right accordingly" (cited in Kong, 2009, pp. 124f). On yet another occasion, he simply stated that the cooperation was a "necessary gamble worth taking" (2430808004, p. 2). Abdul Gaffar Peang-Meth, in contrast, gave a quite pragmatic and some might even say inconsistent explanation: "On the one hand, the raison d'être of this non-Communist resistance movement originally had been to oppose the Khmer Rouge regime. On the other hand, the Front needed aid and assistance to carry out this struggle" (Peang-Meth, 1990, p. 178). In the end, however, the strategic coalition was mainly born out of political pressure: The non-Communists could only count on reliable support by major states if they cooperated with the Khmer Rouge. Therefore, putting the blame on the leadership of the KPNLF and Funcinpec was easy, but in the end, of course, they would have preferred to be supplied with food for their camps and weapons for their troops without relying on the Khmer Rouge (GS-KP1).

Due to the composition of the KPNLF with many former officials from the Khmer Republic, the cooperation of certain members on each side was similarly difficult. Sihanouk's son Ranariddh, for example, once accused the KPNLF of being a direct successor of the Khmer Republic, which is why he would fire all 'Republicans' if ANS eventually won during upcoming elections. KPNLF member Kong Thann commented on this with an analysis of deep-rooted fear of 'intellectuals' in the royalty, thereby also displaying his very own anti-royal sentiments:

> To this shooting pain, Gen. Sak Sutsakhan wept his tears out at a quiet place. […] Kings don't want anybody clever or better than them. If anyone was better, most kings in history usually found ways for fast destruction. This has been the cause for factionalism in Khmer society. (Kong, 2009, p. 117)

Leadership Quarrels: KPNLF

Inside the KPNLF, many different leadership groups tried to define dominant practices and institutions to lead and discipline soldiers. During the early years, more and more Khmer politicians, military leaders, and intellectuals from the Diaspora communities in the US and France returned from exile in order to help at the frontline. Most politicians and intellectuals arriving from the interior, in contrast, preferred to be resettled in the US and France as soon as possible after experiencing the Khmer Rouge and subsequent occupation by the Vietnamese. They had had enough of war and sacrifice and were seeking a quick transit. Only a few stayed behind to fight (or more precisely, to lead fighting). Diaspora and local intellectuals, members of the old military elite, and politicians felt it was their natural vocation to lead the scattered anti-Communist movements. Yet the Sereikar leaders who were active at the border since 1975 or even earlier greeted the new arrivals with skepticism and resentment. For them, the Diaspora leaders in particular simply enjoyed life abroad during the years of terror under the Khmer Rouge and then joined the movement to reap the fruits of their labor as "arm-chair heroes" (Heder, 1980, p. 87). Among the leaders who joined from US was the former Minister of Defense, General Sak Sutsakhan, who formed an alliance with Dien Del in order to change the direction of the guerrilla war but was met with resistance by many of the Sereikar commanders. Infighting among the leaders intensified, and after the murder of his associate Sien Sam On, known as Lert, Dien Del temporarily stepped down from his position as commander-in-chief.

However, it was not only the split between long-serving military and political leaders from the country's elite and local 'warlords' who ruled over mini-fiefdoms along the border that paralyzed the KPNLF. A second major rift existed among the Diaspora leaders due to the anti-royalist stance of Son Sann, who refused to cooperate with Sihanouk and his newly formed resistance group (or vice versa). Old rather personal animosities from the Republican regime that toppled Sihanouk were reiterated, especially because Son Sann chose the foundation date of the KPNLF (by accident or not) to coincide with the proclamation of the Khmer Republic in 1970. After losses in a major dry-season attack by the Vietnamese during 1984–1985, the conflict came to the fore with the formation of the *Provisional Central Committee of Salvation* consisting of Sak Sutsakhan, Dien Del, Abdul Gaffar Peang-Meth, and Huy Kanthoul, who demanded Son Sann's abdication and a closer collaboration with the royalists. The group succeeded in toppling Son Sann from his military command, making him a solely political representative, and forming a *Joint Military Command* with the Sihanoukian Armed Forces (cf. Etcheson, 1987, p. 201).

But still, real military cooperation remained rare, and the 'Joint Command' seems to have never actually met. Some individuals even favored a complete merger with the Sihanoukian forces, seeing a chance for a real united front between former archenemies. But Son Sann in particular was not interested in merging with royalist factions. Son Soubert, the son of Son Sann, however, had his own theory regarding why the quarrel happened: because of the 'CIA' and US interests.[3] In an interview with the author, he stressed that the US became 'deeply involved' in the conflict after 1985 and tried to merge KPNLF and Funcinpec as a means of gaining better control of both. The 'Joint Command' was the US's idea,

---

[3]   For a comprehensive account on CIA operations during the Cambodian wars, see Conboy (2013).

and they tried to lobby against Son Sann and even prevented him from visiting 'his' camps (Interview Son Soubert). However, it is clear that neither Son Sann nor the royalists and the 'dissidents' in the KPNLF were willing to cooperate. Even if the US did influence the decision for a 'Joint Command', they could have easily picked up and make use of already existing infightings and animosities.

The movement became more and more paralyzed during the end of the 1980s and suffered from a high rate of defections among its ranks. Military operations to the interior became almost nonexistent (cf. Bekaert, 1997, pp. 214–17). Commanders stayed back in Bangkok, waiting for the results of the leadership quarrel, and refrained from visiting the camps or going to battles of any sort. The heroic fighting to win the country back was deteriorating, and much of the fighting was to defend what was left of the resistance as a force. While the KPNLAF had a force of 10,000 to 12,000 men before the major Vietnamese attack in 1984–1985, they lost many troops during battles and even considerably more during the leadership quarrel. Towards the end of the 1980s, the number decreased to only about 1,000 men. In the end, however, all estimates were blurred since the KPNLF stated it had about 47,000 soldiers during disarmament following the peace agreement in 1991, with 'observers' believing they had even fewer than 1,000 (Bekaert, 1991, p. 841).

Overview: Habitus Groups in the Field

The Cambodian insurgency comprises several habitus groups on different levels of the three military organizations. For starters, the following diagram gives a brief overview over the military ranks within the three factions:

Table 3.1   Formal military ranks in the resistance according to action

	Khmer Rouge (NADK)	KPNLF (KPNLAF)	Funcinpec (ANS)
Leadership	Party Central Commitee (Military Directorate)	General staff Brigade Independant regiments	General staff Brigade Independant regiments
Mid-range operators	Brigade Regiment Battalion	Regiment Battalion Company	Regiment Battalion Company
Rank-and-file	Company Section (Platoon) Group leader Foot soldier	Section (Platoon) Group leader Foot soldier	Section (Platoon) Group leader Foot soldier

In theory, a group consisted of three soldiers, a section of three groups, a company of three sections – until reaching the level of the brigade. Within the Communist system of the Khmer Rouge, the party's political leadership, the Central Committee, was also the military leadership. In 'war communism', both fall into one, which is why the Central Committee was renamed the 'Military Directorate' from

1981 onwards. Within the Khmer Rouge, military leadership applied to this group, but brigade and regiment commanders were already operators rather than leaders. Within the much smaller KPNLAF and ANS, some regiment commanders could be viewed as members of the military leadership due to their power to influence practices within the guerrilla force. Moreover, there were other, independent regiments and commando groups as shown in Table 3.1 above. These independent regiments were specialized commando units or their commanders defied complete subjugation under the chain of command despite having a relatively small force. Mid-range operators were those who had to implement the plans from the top command but still did not engage in combat themselves and did not know all of their subordinates personally. This was mostly still the case for all small-scale leaders within lower ranks. In the end, of course, this delineation is artificial but also reflects divisions in habitus. Mid-range operators, for example, state that they either only implemented commands from above or that they were "advisors" rather than "commanders" (e.g., BC-F4). Furthermore, its artificial nature lies in the simple fact that a guerrilla force does not fit a formal military organization. Labels did not fit, and many ignored them, especially when in control over a camp with a sufficient number of soldiers to keep them afloat. Some commanders showed "near-perpetual insubordination" to the chain of command (Conboy, 2013, p. 172) and therefore were able to shape power practices within their followership without much interference from outside.

Within the ANS and KPNLAF, at least formally, many brigade commanders also had a position within one of the five bureaus of the army. These five bureaus were in charge of organizing military operations, as one brigade commander explained:

> There were a few offices in the military cluster. The first office dealt with military forces. If I wanted to prepare soldiers for the fight, I would go and ask the first office. We wanted to know how many soldiers there were and they would tell there were one hundred and fifty, but this amount was sick, this amount was absent, and the remaining amount was this much. The second office dealt with information gathering. We could ask this office how strong or how weak our enemy at this particular place was. The third office made the plan. They strategized the fighting. The fourth office organized food, weapons, and the like, and the fifth office dealt with psychological stuff. We obtained information from the five offices and then we could start our operation. (BC-F4)

Before looking closely at the field, a few explanatory words shall be provided about the following Table 3.2. Of course, it does not claim to mirror all possible groups in the field, only those that are constructed by the researcher according to his interviews. However, this table remains what it is: a scientific construction. Moreover, there is one influential group missing, namely, the 'warrior princes' on top of the military command of the ANS. Since there are just two (though highly influential) warrior princes, they do not form a group in a strict sense. Nevertheless, their habitus is very different from the rest of the leadership ranks. In the end, since there are just two, confidentiality would be an additional problem if using material from interviews. Additionally, the anti-intellectual intellectuals at the top of the NADK leadership are widely known, but their habitus, discourse, and practice had to be established by other means than personal interviews. The simple reason for this is that they are either dead or in prison facing an international trial. Materials here had to be collected from various other sources. The study does not claim to be exhaustive. Yet another group that is missing is that comprised of long-serving commanders from the Khmer Rouge, who are either dead (e.g., Ta Mok and Ke Pauk) or, due to the Khmer Rouge Tribunal, fear legal consequences when talking to foreigners. In terms of hierarchy, they would be between the anti-intellectual intellectuals and blank-page leaders. During the Khmer Rouge reign, these former monks and anti-Colonial insurgents were in charge of organizing internal purges. Before starting with the top command, Table 3.2 gives an overview over the field:

Table 3.2   Habitus groups within the field

Military Rank		
Leadership	**Operators**	**Rank-and-file**
Guerrilla strongmen	Loyal tacticians	Pragmatic clients
Intellectual commanders		
Old military elite	Battle-hardened roughnecks	Ethical spiritualists
Warrior princes		
Anti-intellectual intellectuals	Blank page operators	Blank page warriors

Chapter 4
Military Leadership

Within the field of resistance, each group has one major symbolic resource structuring their discourse and practices. This main resource is born out of their social background and emerges from the courses their lives take. Each statement and practice is interwoven with and refers to this resource, which is used to legitimize a power claim within the field. For example, for the intellectuals it is education. The divide between the educated and the uneducated governs whatever they say and do. This main resource is characterized by its exclusive claim by the particular habitus group: They claim to be in sole possession of it, thereby also legitimating their field position and using the resource as symbolic wager within their field struggles.

Moreover, before starting the leadership section with the 'guerrilla strongmen', a small terminological note on the usage of the word 'discipline' needs to be added. Especially in our case, this term can mean two things: First, discipline with regard to good or 'correct' conduct; and second, discipline as obedience to commands and commanders. Although both terms are highly interrelated, the content and its context differ. Especially in interviews, the respondents did use them differently in order to mark a difference between leadership styles. Thereby, for example, it becomes possible within the discourse of some respondents to have extremely obedient soldiers showing poor conduct (due to having a poor leader). Being disciplined in terms of obedience does not mean that you are disciplined in terms of conduct. Therefore, I will separate the two as much as possible by using either the term 'obedience' or 'good conduct'. 'Discipline' will only be used if this double meaning is included and intended. In the Khmer language, this is quite similar with the word *viney* and a differentiation between 'having discipline' (*mian viney*), as self-discipline following the code of conduct, and 'respect discipline' (*korup viney*), as following the rules, being obedient.

Guerrilla Strongmen[1]

For many, resistance had already begun under and against the Khmer Rouge. Remnants of the Lon Nol police, lower political administration, and military escaped execution after persevering in hiding for a while, then fighting their way to the border. For most of these men who went to the Thai border to evade Khmer

[1]   The sections Guerrilla Strongmen and Intellectual Commanders are amended and extended versions of 'Analysing the Cambodian Insurgency as a Social Field', published in *Small Wars & Insurgencies* 25 (2). Copyright (2014) by Taylor & Francis. Reproduced with permission of Taylor & Francis via Copyright Clearance Center.

Rouge units and to seek refuge with Thai patrons and sponsors, or in Thai border military or police units, basic survival was their primary interest and symbolized their strength. The first habitus group from the leadership of the Cambodian resistance are these guerrilla strongmen who built up a position of power while resisting the Khmer Rouge – power that helped them to secure position in the coalition now fighting with instead of against the Khmer Rouge. To begin, the habitus formation will be discussed: family of origin, education, lifecourse information, accumulated resources, and participation in different social fields. This forms the conditions of access to the field of resistance for the strongmen – although the word 'access' is a bit imprecise, since most of them were founders of armed groups, consisting of five to ten men at the beginning, and 'merged' with others to establish the resistance forces. Within the CGDK, the strongmen are the only military leaders who came from a lower social milieu. Battle experience, invulnerability destined by fate, and proven bravery are therefore propagated as the main and most valuable resources in the military field. However, resources gained by strongmen are highly volatile and depend on a credible usage of violence.

Habitus Formation and Lifecourse

All strongmen came from rural farmer families. They share the same background and the same reason for attending a high school despite having had peasants as parents during the 1960s: a massive widening of the educational sector by then King Norodom Sihanouk (cf. Ayres, 2000). Right before Sihanouk was ousted from office in 1970, the Cambodian educational sector increased to nine universities, 3,200 primary schools, and 161 secondary schools – an estimated increase of more than 130 per cent compared to 1954 (Procknow, 2011, p. 116). While during the 1950s a high school degree (Bac II)[2] was highly prestigious and limited to the elite, the value of degrees went down during the 1960s as more and more people graduated from schools (and some universities) throughout the country. However, for the first time even rural families had the opportunity to send one of their children to schools in an urban area. The prerequisite for this was the ability to afford 'losing' one child in fieldwork and the proximity of an urban center with a school they could attend. Most could send only one child: "Our village was very far from the school, so in the family there must be those who studied and those who stayed at home and helped the parents" (BC-F2). Due to the problematic access of rural children to schooling during these years, many individuals from this milieu were quite old when they graduated, mostly due to attending school and at the same time working at home. During the harvest, school attendance went down considerably.

A family's economic position in the subsistence economy of a village can be 'measured' by whether or not they could afford to send one or more children to school. Therefore, these families were not the poorest among the peasantry; rather,

---

[2]   Cambodia adopted the French Baccalaureate with Bac I for general knowledge and Bac II for specialized subjects and graduation after twelve years.

they were slightly better off compared to others in their villages, who needed every hand at home doing the fieldwork. In the end, many rural children faced problems either in completing schooling (dropping out early) or in finding a job later on. However, this was not just the case for the rural population, due to their age or background, but also for higher social milieus, who faced the realization that Sihanouk had created a massive education system but no jobs for the graduates coming from these institutions. However, for the graduates with a rural background, the situation was even worse, since they did not have many opportunities at their disposal. Without social or economic resources, their acquired cultural capital was almost worthless, and they could not afford to wait for improvements. Hence, other than going back to the villages to become farmers again, one of the best opportunities after the coup against Sihanouk in 1970 was to become a soldier in the military under Lon Nol to fight the Khmer Rouge, to join the police, or to become a low-ranking official (e.g., a village or district chief). Due to the war against the Khmer Rouge forces, the government was recruiting many young graduates and pupils. The wave of graduates as produced by the Sihanoukian reforms in the educational sector created a large pool of possible recruits for the state military. However, due to their background, the entry level in these fields was relatively low and, with regard to the military, remained at the level of the rank and file. The gains they saw in terms of education did not result in upward mobility in Cambodian society. Instead, their low status was only transferred into a different social field, thereby resulting in merely horizontal mobility. Instead of plowing the fields, they now joined the forces to defend the cities from being taken over by the Communists at the lowest level and fought in the front line.

Khmer Rouge Genocide: The Unlikelihood of Survival

On the very first day that the Khmer Rouge took power and captured the capital Phnom Penh, they expelled all city dwellers to the countryside. While forcing everyone, including the elderly and sick, to leave immediately and killing those who refused, Lon Nol officials, policemen, and soldiers were singled out for execution. The state had to be 'cleansed' of the old system and its oppressors (cf. Kiernan, 1996; Becker, 1998; Chandler, 2008). During the earliest wave of purges, members of the Lon Nol political and military apparatus were the first on the lists to be (in Khmer Rouge terminology) 'smashed'. While the top leadership of the Lon Nol apparatus largely made use of opportunities to leave the country in the days and hours before the Communist troops reached Phnom Penh, the lower ranking personnel did not have the chance to leave the country. The only thing they could do was to take off their uniforms and hide among the expelled population, pretending to be 'ordinary citizens'.

One of the most important social divisions introduced by the Khmer Rouge regime was that between 'new people' (or '17th April people') and 'old people'. While the regime officially divided the Cambodian society into twenty social categories, in practice inside the cooperatives it became either a tripartite categorization of people with full rights, candidates, and deportees. However, the basic division for many cadres and victims of the regime was one between 'new people' and 'old people'. 'New people' were those who lived in the cities before 1975, had a high level of education, and/or came from upper 'capitalist' and 'feudal' classes, while 'old people' were those who lived under Khmer Rouge control in the countryside before 1975 and had a rural and mostly impoverished peasant background (ECCC, 2010, p. 80). In the village, due to their military background, the strongmen had to conceal their identity. Some succeeded in doing so, but most of their colleagues did not and were arrested and sent to one of the almost 200 prisons, eventually ending up being executed on a 'killing field'. During their stay in the countryside, they did what most of the other people were doing: working on rice fields, constructing dams for the nation-wide irrigation system, and being constantly under surveillance. Due to a lack of food, being overworked, and poor medical conditions, many died over the years. As former city dwellers and 'new people', they were the last on the list to receive food or medical attention and first to be imprisoned or even executed for any minor wrongdoing (Ebihara, 1990, p. 25).

Some smaller units managed to fight their way to the border right after the Communists took power or remained active in the interior for a while, but these units had been decimated over the years by Khmer Rouge raids. Over the following years, they received a constant but small flow of survivors of the state apparatus, whose identities had been revealed and who fought their way to the border to look for a resistance – and a better chance for survival. Yet, in general, the state of the insurgency, active via some local Thai police and military forces without wider attention and funding, remained rather desperate. The journey to the border plays an important role in the interviews and in the life stories of the strongmen as it – in their own view – proves their strength and, for most of them, even their sense of being destined to survive. While the strongmen describe the years under the Khmer Rouge rather concisely with the lack of food and the threat by cadres to take them away, they describe their flight extensively to the very last detail of its hardships, the dangers of being killed while passing Khmer Rouge units, the luck they had, which was guided by Buddha's interventions and securing hand, and their repeated near-death experiences along the way. The whole journey is described as "a single piece of luck in a thousand dangers" (BC-4F). Only fate, magical protection, and 'Buddha' secured a way out when they met soldiers with weapons or were shot at with 'hundreds of bullets' and 'mysteriously' survived. Survival is not a matter of mere accidental luck nor of special skills; rather, luck is a matter of being chosen to survive.

But first they had to survive their years at the Thai border. Many who were not 'chosen' either died over the years in cross-border raids by the Khmer Rouge or led a desperate life in Thai prisons or as mere intelligence workers for the Thai border police. Although they commanded only few people with even fewer weapons, some even attacked each other in a tense situation with only few possibilities for finding a spot to make a 'living' and defend themselves against raids from all sides. But in the end, these scattered armed groups formed the basis of the resistance movement along the border, to which many Thai-Cambodian people gave their support and tried to lobby help for from the outside and from the Diaspora. While local and

Diaspora activists could not gather support for a war against the Khmer Rouge, the situation changed when they lobbied for fighting against the Soviet-backed Vietnamese, who expelled the Khmer Rouge from the country to the Thai border and thereby instigated a flow of refugees settling close to the Thai territory and asking for assistance. Then, the armed groups were in charge of 'defending' those encampments and the populace seeking shelter, food, and resettlement abroad. But simultaneously they 'defended' their sudden sovereignty over people, goods, and money and showed perpetual insubordination to the top command.

Access to the Field: Strongmen Becoming Powerful

Having a couple of men, weapons, and the capability to fight was a valuable commodity, which put people in powerful positions among all of the desperate refugees. Furthermore, when people from abroad (politicians, intellectuals, military leaders) joined in, these 'forces' could be used to convince major patrons that there was already a resistance on which they could rely and in which they could invest. And on top of it all, those forces had – contrary to many others along the border – a proven non-Communist record. Although each former general and each politician agreed that they definitely needed a polish in discipline and more clear-sighted goals to be a respectable force, they were something to start with and, in the end, the closest thing they had to a 'resistance force'.

When 'joining' the front, they had almost nothing in hand: no social resources in terms of contacts to the top leadership, no prior economic resources, and a comparatively limited education. But they quickly made a small fortune along the border with black-marketeering (although this had to be largely redistributed to sustain their followership), and they received formal recognition as brigade or regimental commanders in the KPNLAF or in the Sihanoukian army (ANS). Only a few people who had followed them from the early days along the border or even earlier during their flight stayed close to them, and trust was a central commodity that they guarded closely. From their own perspective, the most valuable thing they had and therefore promoted in the field was having been directly involved in many battles and having proven bravery and strength as survivors of countless dangers. In short, their capabilities and biography as a low-ranking soldiers endowed with fateful strengths is the reason for their superiority. These symbolic resources, however, only made sense within the field and especially due to the breakdown of large parts of the former state militaries and their leadership. And it only made sense during the brief period when the field came into existence when there were, at times, just a few soldiers resisting along the border.

While their high leadership position within ANS and KPNLAF was facilitated by their access to profitable routes along the border, their threatened and shaky position in the field was largely enabled, legitimized, and symbolically secured by a discourse on their soldierly and magical powers received by fate. However, strongmen face problems to secure their sovereignty due to the volatility of their symbolic resources and status within the field. But as long as they remain 'untouched', they believed, they are chosen by fate to be beyond any volatility.

Classification: An Unequal Distribution of Fate

For the strongmen, it is fate which determines a person's capacity to gather strength and become powerful. The guerrilla strongmen classify their ascendency from an ordinary Lon Nol soldier with a rural background and limited education to a brigade commander as a history of bravery and an unfolding of fate. By combining the Buddhist notions of fate and of collected merits through proven soldierly conduct such as bravery, the strongmen explain their rise from 'ordinary' peasant families up to powerful commanders and later on, as in our case here, even to Acting Heads of State:

> People's fate is not always the same. We may also say that it is fate, which determines someone's power. Some people tried very hard, but could not become a commander. But for me, I never expected to be in a leadership position before because I was just a rural farmer and didn't receive any higher education. However, due to my efforts together with my predefined fate, I was able to succeed. […] Let me say something funny here. I am a son of farmers, but I was able to become a leader. It was my fate to be an Acting Head of State on behalf of the king while Cambodia was facing a political dilemma […]. There, I was able to perform king's tasks for the country. But if we look at Ranariddh, although he is the son of the king, he never actually had the fate to be an Acting Head of State. This was fate ((laughs)). (BC-F4)

Fate is unequally distributed among men. Not everyone can be a leader. But at the same time, fate is not enough; it has to be fulfilled by collected merits and, in the case of military commanders, by proven bravery and success in actual combat. Devotion to soldierly conduct and rules makes soldiers grow in strength within the boundaries set by their individual fates. Bravery as proven in many battles and by surviving the Khmer Rouge genocide is the reason why many of them became commanders: "Commanders were chosen based on their bravery and collected experience first". And fate makes a commander capable of accomplishing many soldierly merits and gaining power, which is why they became commanders while other brave warriors did not rise to leadership positions: "I was meant to be commander" (BC-F1). He is distinguished through outstanding military deeds, ascetic conduct following spiritual codes, and accomplishments in battle.

Role Modeling in Action: Obedience and Viral Spread of Conduct
Being a role model, or fulfilling a physical ideal, makes a leader strong and, in turn, sets rules of conduct for the leader and his subordinates. Being a model disciplines the commander, who has to stick to soldierly ethics himself:

> The leader himself had to be obedient in order to be a model for others. […] For me, I had to stick with good morality for many years. I could not do anything bad. The most important thing was our sacrifice, not being self-centered. (BC-F2)

Sacrifice and devotion to a model is what the strongmen expects from everyone, no matter which rank or position in the group. The commander, therefore, is not essentially different from an ordinary rank-and-file soldier. He is 'only' stronger than his subordinates and chosen to lead due to his superior soldierly conduct making him capable to acquire power:

> We had to make sure that everyone is morally high. This could be realized only if we tried very hard. Military rule says, 'When soldiers sleep, the commander sits. When soldiers sit, the commander stands. When soldiers stand, the commander walks and runs'. We all adhered to this rule. Each commander, including me, had to be hard working. We had to perform more difficult tasks than our subordinates. Being an example was truly important. (BC-F3)

All soldiers have to follow a soldierly ideal with its set of rules to become stronger. If they follow that ideal, they might gain strength; if not, they will certainly fail, stay behind, or might even be killed in combat because fate did not plan for their survival. However, by being a powerful role model, a commander demonstrates his conduct to his subordinates, who always mirror his behavior. Thereby he risks disseminating poor conduct as well: "If the commander himself wasn't a good example for his followers, the soldiers would do bad things. This was an important thing" (BC-F2). The commander's power can easily be used negatively if he himself deviates from the soldierly model of obedience and bravery. In practice, this means that if he is afraid, his followers will be as well. If he does not show loyalty, his followers become disobedient as well. Indiscipline on side of the commander results in a behavioral domino effect, spreading virally to all his subordinates. Since he is the originator of soldierly conduct, the commander had to be a monk in mentality:

> I was different from others because I sacrificed a lot for soldiers. Human beings are very greedy. Although I didn't have my hair shaved and didn't put on monk robe, I was a monk mentally. I always joke about this; for me, diamonds are just the fragments of a broken glass. Diamonds were not important to me at all. I never intended to get money from others. When we become a monk mentally, we have to follow this firmly. But people are usually very greedy. I won over others because of my sacrificial leadership style. I was also an example for soldiers. Before blaming others, we had to make sure that we ourselves were good first. So we had to sacrifice and be a model for others. (BC-F4)

The commander is the spiritual center of soldierly conduct and under pressure to be obedient to the ideal like his subordinates should be to him.

Spiritual Discipline

> We also relied on magical beliefs. We drank medicine, tattooed our bodies, and so on. When everyone soaked themselves with impenetrable medicine, they seldom got injured. This made them mentally strong. Nobody was afraid of fighting in battle because their skin was bullet-proof. Tattooing was also

believed to be very magnificent. Its effectiveness was proven in practice. They didn't get injured although they had stepped on a mine. (BC-F4)

Every strongman was tattooed all over his body and wore amulets and other protective items. Spiritual and magical powers were constantly stressed as being part of an ideal soldierly conduct and force to collect strength and power. There were various items in use during the years of civil war. For example, solid pig spurs, whose value lies in their scarcity:

> I also used a solid pig spur. The solid pig spur was very rare. Pig spurs are usually not solid. Of hundreds or thousands of wild pigs, there is only one with a solid spur and which can see things clearly at night, and cannot be shot. We attempted to shoot it several times, but we could not shoot it successfully. It was a very effective item. (BC-F4)

In addition to bones, spurs, and handkerchiefs there were various magical drinks and medicines as provided by the traditional *kru khmai*. But these magical items had one major hitch: their effectiveness was unequally distributed among the men. The magical power and effectiveness of such items are correlated with the individual's fate:

> It was not that everyone was helped by that magical power. Only one or two among all were actually helped by it. This was one point. Second, what everybody did was drink and soak themselves with impenetrable medicine. Only this could help everyone in the combat. But for the magical power, not everyone was helped. Out of one hundred people, only one person was helped. This was relevant to the first problem, because we had faith in it. Just like me, I had faith in magical instruments such as solid pig spur or the use of *Keatha*. I don't know how to explain it to you, but to me, I really believed in that because I saw that I really was able to survive because of those during several times of great danger. (BC-F5)

All these items, however, develop their full potential only in accordance with correct spiritual conduct. For some, strength is volatile since they are not chosen; for others, it is not volatile, and they are capable of gathering ever more strength. However, it was not simply the use of magical items that made soldiers strong; behind such items was a whole *system of conduct and spiritual rules* guiding the correct use of spiritual powers. These rules that should be followed were decisive and ascribed even a commander's eye with clairvoyant powers:

> We had to chant something, and we didn't just chant before combat. We chanted regularly, such as every morning and again before we went to sleep. As I kept on chanting, I was able to be pre-informed if something was going to happen in combat. The pre-information was submitted through my right eye's wincing. If it winced slightly, it meant the upcoming problem was just a minor one. Similarly, if it winced strongly, it meant the problem will be so big that it could even lead to the death of my subordinates. As I knew things early due the wincing, I had to be careful. (BC-F2)

Potency comes from the devotion to rules – rules that are not only those rather classical soldierly codes of conduct, such as obedience but also are spiritually and animistically guided. There are various seemingly arbitrary religious codes and rituals that serve as disciplinary techniques and an explanatory framework for the difference between those who survive a battle and those who do not. If one misreads or ignores the rules as set by the magician or *kru khmai*, these protective forces might cease in battle – with the result that the warrior might get killed. The rules that come with the use of magical items can be rather technical rules related to diet or hygiene, like not to eat pork or not to wear your protective handkerchief while on the toilet, or more strict and ethical rules, like not to yell at your wife or subordinates (Maloy, 2010). But these rules could differ between different *kru khmai*, who were more or less strict as one strongman pointed out: "Generally, we were prohibited from sleeping with other people's wives. However, my *kru* said something different. He said that raping others would be definitely wrong. But if both sides agree [to make love], it was fine. My *kru* said so" (BC-F5). Thus battle success and thereby also survival is a mixture of obedience to (spiritual) rules, a model set by the commander, accumulated strength, and fate. Within the refugee camps, various monks and *kru khmai* helped soldiers to become better warriors, provided items, and gave advice and training. Some monks even took part in making soldiers shoot effectively and survive the battle despite Buddhist rules forbidding participation in violent acts.

Excursus: The Monk and Non-Violent Participation
It seems a short digression is necessary here, although Buddhist organizations in camps might be slightly off topic. But the reference to Buddhism and violence calls for an explanation, and the context should be clarified. Although I do not intend to discuss 'real vs. misinterpreted' Buddhism, the simple fact remains that there seem to have been some monks (beyond magicians such as the *kru khmai*) who provided teachings and even trainings to soldiers. These teachings and trainings had a lesser importance and were less prevalent than those of the magicians who provided spells for animistic protection or items with spiritual powers. However, there was a 'Buddhist Association' in KPNLF camps. Monks and other members of the association are important for this study in two regards. First, they believed they could discipline abusive camp leaders; and second, they provided training *to make soldiers fight*.

The first point is easily described and not as controversial as the second. The association sent monks to unruly or undisciplined camp leaders to make them behave 'better'. The Buddhist morale training was aimed "to reduce savagery, barbarity, and violence of the undisciplined power holders and arm bearers in the no man's land" (Kong, 2009, p. 171; slightly corrected). Many camp leaders and brigade commanders were given a monk as companion to make them drink, gamble, and kill less. Whether or not they were successful is impossible to track, but these Buddhist 'advisors' believe that they were (e.g., MBA). For most people, however, the second point is much more troubling. When hearing that they trained soldiers to fight, one thing immediately comes to the reader's mind: Buddhism in known for its central doctrine of non-violence. True, but there is always a way around, and monks in the camps, as did most other people, saw the necessity of saving their own nation from being swallowed by Vietnam. However, the monks themselves were not allowed to cause any harm or to attend anything harmful. The head monk of the association gave an allegory to explain the logic behind the monks' strategy that enabled them to do what they 'normally' should not do:

> R: Let me give you an example. A monk has seen a thief running right in front of him. Later, somebody comes and asks if he has seen the thief. If he says he has seen him, it seems like he is encouraging the police to arrest the thief. Yet, if he tells the police that he hasn't, he would be telling a lie. Both answers are wrong. So, what can the monk do to get himself out of this? When the monk saw the thief, he was standing here. If he just moves back a bit, and changes his previous position, he could tell the police: 'While I was standing here, I did not see the thief'. ((Laughs)) When the monk saw the thief, he was standing here. But when he moved here, he did not see him. So this is a win-win situation. The monk neither told a lie, nor has he told the police to arrest the thief. We did not intervene in the two's affairs. You get it? When the monk saw the thief, he was standing here. If he just moves one step away, he could tell that he did not see him while standing here ((laughs)).

> I: I am a bit doubtful and I would like to go back a bit to clarify something. The monk INTENTIONALLY moved a step in order to change his standing position?

> R: He did in order to survive. We do not want our words to have any bad impacts on others. Just think about this example. And you don't have to wonder whether monks taught soldiers or not. (HM)

The morality of Buddhism, it seems, does allow digression from its rules and bending them slightly for practical purposes. The head monk even defends the right to kill in self defense and stresses that this is not sinful since it is not an intentional act and the situation 'exceptional':

> Buddhism prohibits killing (4) but this case is exceptional. If others invade us, we have the duty of self-defense. We did not INTEND to do it though. This is not a sin. I want to clarify because in case you are curious how we could defend our territory. If foreigners come, we could not give them our land because it's our heritage. (HM)

The head monk connects the Buddhist idea about intentional acts and sin: The one who kills without intending to cannot act sinfully. It is as if someone scrunches an ant and cannot be blamed for killing it. Of course, this comparison is a bit far-fetched, but for the monk, the same principle of intent is at work. But there still remains the central point that monks are not allowed to train soldiers and to take part in harmful action. However, this does not mean that they did not train soldiers to fight, as the head monk already suggested. They made use of a quite simple trick, reflecting the allegory on the thief: When the soldiers came to the pagoda dressed as 'civilians', they would be trained in several techniques while simply not talking about the military purpose and the military background of those attending the sessions. Officially, there is no intent to train them for battles, just for 'matters of life' in general. Kong Thann, member of the 'Buddhist Association', which was in charge of pagodas in several camps, explains the content and background of the Buddhist military teaching called the 'three Vedas': Trainees would receive three skills, namely, martial arts for making war, horoscoping for predicting occurrence of and situations in war, and protective charms (Kong, 2009, pp. 77–9). These techniques mainly aim at lowering the fear and reluctance of soldiers to go to combat. And they are combined with certain rules as well as mental and physical exercises. Thereby a soldier is supposed to get strong enough for a battle, both physically and mentally, and, as a former member of the association said, to focus on shooting:

> [In his teachings] Buddha explains that soldiers have to first shoot straight, second shoot far, and third shoot right on target and defeat the enemy. This is all he explains. Before going to the battle, soldiers have to first obtain clear information about the combat. This is what Buddha says. Before breaking the target, one has to do an extensive study of the battle. If we are too close to the enemy, we should back up a bit. When we stay far from the enemy, we use instruments to shoot from the distance. There were sets of instruments for soldiers to use. Buddha knows everything. It's just that he doesn't go into detail. (MBA)

While monks only seldom actually trained soldiers to focus on their job, to meditate, or the like, many monks provided protective items and chants before going to battle as well. For the most part, they gave some advice as the head monk indicates, and then simply avoided going into detail like he did it as well. However, as protective service, for example, they made handkerchiefs called *yorn* or other items with magical powers to safeguard the warriors. However, despite having some magical powers, a commander still had to *patronize* his followers in order to make them loyal subordinates.

Caring Patron and the Trusted Entourage

> I was nasty. I hit them [his soldiers] whenever I needed to. Yet, they still loved me. It was because I was honest. I never kept money for myself. [...] Some did not care about their soldiers when getting money. This made their followers unhappy. [...] I never exploited my soldiers financially. We ate what we had together. I always ate with my soldiers. We went everywhere together. I never hid anything from them, and I always gave them something. I always gave them money or anything they needed. (RC-KP2)

The strongman is the center of arbitrary nastiness but also of goodwill as a caring patron. No matter how powerful a strongman might be, the loyalty of his subordinated soldiers does not come naturally. Power here is a personal quality and does not necessarily result in a structural position with control over many people. A lonesome strongman in the woods can still be powerful due to his accumulated strength. Although the commander might be a strong model, he still had to earn his followership by taking care of his subordinates, thereby reiterating classic sociocultural schemes of patronage as a mode of rule: "A commander had to know his soldiers' happiness and sufferings" (BC-F1). First and foremost, a commander had to secure his followers' basic subsistence, listen to their needs and worries, allocate resources to them, show them that he is one of them, and be humble (BC-F2; BC-F4). Being 'humble' means that his domination is bracketed, and the strongmen feigns fraternity and equality. As a potent center, the patron's power and conduct protects and takes care of his clients. The patron takes care of his subordinates and provides them with everything they need (wealth, shelter, protection), while he, in turn, demands unconditional loyalty and service.

But as Alex Hinton emphasized, these networks suffer from constant uncertainty regarding trust and the threat of a breach of loyalty that results in a certain degree of constant unresolvable distrust: "Such personal networks, however, are often filled with uncertainty, negotiation, and distrust" (Hinton 2004, 119). However, especially under the conditions of violence and constant danger, loyalty is always at stake, and the strongmen knew that a 'compliant face in front of the curtain' did not mean that his clients were actually loyal to him in their hearts. Being under constant threat of survival and dangers of all sorts makes loyalty the most important and most insecure quality of fellow soldiers, as raised in a saying often cited by the strongmen: "Different mothers at home, same in the jungle" (e.g., BC-F1). Networks are always and for everyone in the field filled with uncertainty, but the main insecurities for strongmen lie in the volatility of their resources and in the heavy competition between countless groups operating along the border, making their leadership position highly fragile. While in the woods, total devotion and sacrifice for the leader becomes the norm, exactly because of the volatility and fragility of their sovereignty. The strongmen's discourse reflects their struggle for a legitimate leadership position and to retain their status – particularly against members from the military and political elite returning from the Diaspora believed to be taking over 'their business'.

Field Conflicts: Intellectualism
Unlike many others within the non-Communist leadership, strongmen are largely those who stayed in Cambodia during the Khmer Rouge regime. Those coming from abroad to unite the forces and to take over leadership were greeted with a good deal of skepticism. Even being a son of former King Sihanouk like Prince Norodom Ranariddh did not help one to earn respect among the guerrillas and battle-hardened warriors. Although Ranariddh received high honors as a former professor of law, leadership in Cambodia, as one strongman stressed, is much different: "He used to teach law, and France accepted his knowledge. But management principles, strategies, and leadership for Cambodians are very different to those overseas" (RC-F2). Intellectuals and the old military elite were

regarded as mere 'arm-chair heroes' without experience in actual man-to-man combat. In their eyes, actual experience is the key to good leadership:

> Experience was more important. Theory was one thing, but what we actually practiced was more important. [...] First, I did learn a lot from my previous military commanders. But in the end, the most important source of learning was actually from the practical experience in battle. As I have said, we applied what we learned from our previous experiences. (BC-F4)

The strongman's superior capacity for soldierly conduct such as obedience to rules, braveness during battles, and strength makes him the example everyone should follow. Hence, good leadership can be seen in the practiced role model behavior of an experienced strongman, whereby the strongmen draw a clear distinction between themselves and intellectual commanders. "Certificates" play no role in good military leadership – they are for "offices only" (BC-F1). What you need to lead a guerrilla force is actual combat experience, the capability to be strong warrior, and adherence to a spiritual-religious code of conduct supplemented with various magical items and tattoos. Neither education nor military expertise (in soldierly drill, training, or strategic planning for battles as regular commanders would emphasize) is used as a resource in the symbolic struggle against other commanders in the field; it is 'solely' the commander's superior strength as a warrior and a survivor of dangers that demonstrates his power and position.

Power Practices: Sovereignty

The structure of the strongmen's power practices reflects what Michel Foucault calls sovereignty (Foucault, 1977). In sovereign power, the center has a moral role and as a potent center, diffuses good conduct throughout all its fiefdom. The sovereign power's moral right to rule is established and secured in every act of power aiming at an individual relation of loyalty and, if necessary, reconstituted by force and blood (Foucault, 2009, p. 28). The sovereign power demands unconditional and personal subjugation under his rule alone, not that of the state or any formal organization, with organizations like the KPNLF or Funcinpec being utilized at best. All rights to reward and punish are in his hands, thereby making him the ultimate source of justice and goodwill. Each act, each punishment, each reward, each promotion carries the imprint of his sovereignty and his personal will that needs to be secured. His ever-changing and therefore unconstrained will and arbitrary conduct symbolizes his sovereignty and unquestioned rule. There is no law – not even a rule or behavioral pattern – above the sovereign power, and everything and everyone needs to be catered towards him. Sovereignty is shaped by patrimonial patterns, in which the followership is patronized for its loyalty. The main means of control for the strongmen becomes face-to-face enforced group cohesion. However, their sovereignty is constantly under threat, which is why they constantly have to reintroduce their sovereignty by violent and 'ruthless' action.

Patronizing a Followership

The strongman's subordinates were (even spatially) organized in concentric circles according to their degree of loyalty and trustworthiness, with intimates held close and always in sight. A strongman promoted and rewarded only those who were trusted and close, those who proved their loyalty by bravery in many battles for him – be it at combat operations or the 'home front'. They screened for uneducated former peasants, who did not receive much more military training than themselves. Promotion went to those who were 'brave' and 'honest', as they believed in a carry-over effect in mid-range leadership as well since "if a unit had a brave and honest commander, the soldiers in the unit would become strong accordingly" (BC-F5). The strongmen's entourage came from the lowest strata of the military, from the rank and file of the resistance with combat experience, 'proven' bravery, loyalty, and roughness – in strong opposition to those being recruited by other intellectual commanders, who screened for recruits with higher levels of education and disregarded the 'illiterate' rank-and-file soldier as unsuitable for rank promotion.

Therefore, most of the strongmen's followers were quite young when they became soldiers for the first time (starting between the ages of twelve and sixteen). For many of the strongmen's subordinates, it was unconditional loyalty and honesty in exchange for material goods that seemed to count most, and not competence or any sort of skills, which could be rather frustrating for some: "Soldiers in other countries could get promoted based on their level of education and achievements. But promotion wasn't done this way during that time. Why do I say so? Because some people did not even shoot a bullet but could get even higher positions than me, just by being closer to the commander" (GC-KP2). The whole patronage system was financed by gaining control over black markets and by diverting aid flows within the camps. Therefore, many Sereikar warlords were in constant battles for profitable routes, maybe even more than they were in battles against the Vietnamese 'occupiers'. A former commander described the logic behind attacking each other's camps and the excuse found for their actions as follows:

> When commanders saw that a particular camp was conducive to doing business, they would try to convince other leaders to attack the camp. Old Camp, New Camp, and Nong Chan Camp did not go along well with each other as each was trying to compete for the black market. They always accused each other of betraying the front. [...] A camp with a high trade would be accused of betrayal and was attacked later on. (BC-KP1)

Loyalty, not only of the patron's entourage and inner circle, was constantly monitored and was dyadic in face-to-face exchanges. In order to get in touch with a strongman, everyone had to gain the trust of the watchmen in his entourage first and/or gain his trust personally during many talks, meals, and drinks. Even for the researcher, for example, interviews included middlemen and longer passages of informal joking, eating, and drinking before he or his watchmen would allow anything to happen. Building up trust and showing personal affection is not only the key to gaining his trust but also how he himself tries to establish trust among his personnel. For example, all respondents highlighted the need to stay with their subordinates, share everything, and eat with them – in short, to be a good and humble patron. This practice, however, is spreading more fear than affection, as a brigade commander's guard unwittingly pointed out:

> I would give you an example. If there were some fruits on the table, he would give half of it to us. He ate what we ate. He didn't mind me sitting with him. Besides me, none of his staff dared to sit at the same table like him. (BG-KP1)

Ostensible equality between unequal individuals is a power strategy that puts a lot of pressure on the subordinates. By bracketing the power relation, everyone was kept close so that no one could evade the commander, the group, or its cohesion. Group cohesion is the main tool used to control a commander's men and is constantly invoked to secure trust. Promotion does not mean to rise in (military) rank but to gain access to the inner circles of the group. Enforced fraternity based upon a pretended bracketing of superiority and domination instills fear due to a constant surveillance of 'true loyalty' and affection towards the commander. As a reward after a months-long mission to the interior, strongmen mostly organized parties for all of their subordinates (RF-KP2). Here they sat and ate with their subordinates and questioned them on their battle experience and performance. The party is the ultimate symbol of sharing, the norm of group affection as well as the humble devotion of the patron. Having a party and drinks with a strongman was not very relaxing for everyone, but their honesty, complete transparency, devotion, sacrifice, and bravery were constant tributes to the 'group', which the sovereign commander demanded for his role as a caretaker.

Punishments
Giving and sharing is always a sign of the strongman's symbolic superiority demanding loyalty, as is punishment. Formal systems of punishment, which intellectual commanders repeatedly attempted to implement, especially in the KPNLF, did not have a chance due to the basic function of punishment as a personal domain in sovereign power practice. A formal system would have undermined the basis of sovereign rule, where the sovereign figure is the only source of justice. Military formalization or any sort of judicial system would threaten their sovereignty. The exact content of the penalty under a strongman's rule was decided on a case-by-case basis and relates to the rank or 'importance' of the wrongdoer and the question of whether or not he was threatening the authority of his patron. If not, he was simply 'asked' to stop stealing, raping, killing, or whatever he did. As a rule of thumb, for minor wrongdoings, "no punishment was imposed. They [the commanders] would just tell us to stop doing this, and do that instead" (BG-F1). Many subordinates of strongmen simply stated that minor offenses were 'OK' (BG-KP1). Stealing, in particular, was tolerated since it was a comfortable way to finance the unit and allocate resources to the entourage. In the event that it infuriated civilians or other forces within the movement, they were told to do something else instead. However, those threatening the sovereignty of their patron by, for example, refusing his orders or questioning his role as a

Inside Cambodian Insurgency

role model would be killed in retaliation (BC-KP5). Showing disobedience and disrespect of the sovereign commander was the main reason that subordinates were killed: "If someone dared to disrespect, he would be killed. This made others very respectful towards their commander" (RC-F1). For the rank-and-file soldiers, the arbitrariness of punishments – with slight mistakes possibly being punished by death – led to significant fear of strongmen: "Even a slight mistake could result in death. We did not rely on law because guerrilla forces did not have any laws. If we made any mistake, we could be killed" (RF-KP2). You might be killed, but you might not be. When thinking about how his own commander differs from others, the rank-and-file respondent replied: "(4) Some commanders only thought about killing their own soldiers. If soldiers made any mistakes, they would be shot dead. Others at least deliberated before shooting their own soldiers" (RF-KP2). With his leadership under threat, the strongman constantly had to reinforce his sovereign will – a will that is guided by his personal and thereby arbitrary mood alone.

The main problem for the strongman was his fear of a viral carry-over effect among his men, which would undermine his fragile leadership, leaving him a lonesome strongman in the woods. If a subordinate did something threatening order within the unit, "there was a way to deal with it. When they were known to have really made that mistake, they would be killed. We'd rather kill a person rather than letting the rest become spoiled too" (BG-F1). Spoiling the rest of the group was a major danger, which is why, as another strongman said, "we may kill [the offender] because keeping him might spoil the rest of the group" (RF-KP1). Punishment was important to keep sovereignty as a rule but also to prevent a viral spread of indiscipline in terms of disloyal behavior. And since there are no chains, killing offenders in a guerrilla war was almost a necessity for the strongmen: "If they did not change after instructions, we would have to kill them. Guerrilla forces are different from conventional forces. We had no chains for them" (BC-KP5). Although the rule was 'an eye for an eye, a tooth for a tooth', being a patron, who gathered war-troubled and lonesome people, demanded some leniency:

> We simply followed the maxim 'an eye for an eye, a tooth for a tooth', so that soldiers would feel threatened. However, we could not always be hostile to them since they came to live with us because their parents had abandoned them. They regarded us as their parents, so we could not always kill them. We had to decide depending on the particular situation. (BC-KP5)

The good patron shows leniency by deciding from situation to situation whether punishment or good will shall rule, guided only by his very own personal mood and whether he fears a threat to his rule and a derailing carry-over effect, in which his superior position and strength becomes questioned. Being strong, brave, and violent is part of their strategy to earn respect in the field, codify superiority, and secure his fragile sovereignty.

Training
Theoretical or any other kind of formal training like in the KPNLF Cadet School was largely ignored by the battle-hardened strongmen. An ordinary soldier, they

said, should learn from actual combat and not from military drill or any kind of theoretical training. The training at the movement's Cadet School, therefore, "was designed for commanders only. Ordinary soldiers' training took place in actual combat" (BC-KP2). Asked about military trainings, the rank-and-file strongmen stated that they did not receive any training other than occasional basics, such as how to unlock a weapon (e.g., RF-F1; RF-KP1). Most had to rely on themselves, learn by doing, or resort on earlier training: "Fighting in battle was completely based upon soldiers' own techniques. Some simply applied what they learned under the Khmer Rouge regime" (BG-F1). While the newborn non-Communist leadership set up training camps and schools, the strongmen never went there personally to get further training, and rarely sent one of the subordinates to hear some lectures. Similarly, the whole system of political indoctrination that was designed by the political and intellectual leadership, in which the overarching goals of the movement were taught to the rank-and-file soldiers, did not make much sense for the strongmen. For them, there was enough hatred and anger that they could build upon and that simply 'reinforced' everyone's feelings: "Everyone was angry enough". And if not, "we simply told them that the enemy killed their parents" (BC-KP2).

Instead of drills, indoctrination, and other formal modes of training, the strongmen believed that only enforced and visible guidance by a role model in action can prepare soldiers for battles. It was especially helpful if soldiers were too afraid or weak during their first battles. Overcoming fear does not occur via pre-combat training but by mixing weak with strong warriors so that the weaker can learn by their example: "Slowly, we mixed [those who were afraid] with brave soldiers. Sooner or later, they would also become brave". Bravery is, furthermore, directly related to moral strength and discipline: "We explained to them that they were to follow morally strong soldiers. We let them go and after a while, they would also become strong" (BC-F2). Success in battle is a sign of moral strength as well as loyalty to the sovereign patron and commander. Thus, this form of role modeling makes soldiers brave and loyal, who, without any preparation, were simply forced to learn from the example of others in action. Those who failed or died in combat simply did not stick to the commander's model of bravery and obedience or follow spiritual-religious codes of conduct; therefore, they faced their very own fate.

Combat Control
Beyond hosting parties after combat, strongmen also monitored missions on the spot. While other commanders sat back in their offices and demanded written reports and relied on TPO radio transmitters to monitor their subordinates' battle performance, strongmen ignored these bureaucratic techniques that others deemed rather convenient. In the end, the majority of reports did not end up anywhere within the strongmen's units, since they were mainly used by the party's upper general staff to check on units and their respective commanders (BG-F1). The strongmen did not allow others to monitor their own units' performance, battle techniques, economic activities, or leadership style. The chain of command was ostensibly acknowledged, but real interference into sensitive areas of command was strictly inhibited, as were competing penal codes.

Combat operations were monitored closely, and disloyal behavior was punished right away. Battles and camps were surrounded by checkpoints to prevent defections. In case soldiers were too "weak" for fighting, fear was no option for them, simply because there was no way that they could go back (as with some other commanders): "It was impossible to be afraid. They had to go [to the battle]. If someone didn't dare to go, there would be someone at the back to shoot up to the sky to threaten him. Or he could be killed" (BG-F1). Fugitives and defectors were simply shot, and signs of fear were not permitted.

Intellectual Commanders

> I was known to be an intellectual amongst everyone in the camp. The rest were just ordinary refugees. (BC-KP2)

For many intellectuals, especially within the KPNLF, resistance started even earlier than against the Khmer Rouge. Many date their loyalty back to the Democratic Party, competing with Sihanouk and fighting for independence from the French during the 1940s and 1950s. Even though most are much younger, they emphasize their 'ideological' belonging to and high respect for the party that was founded in 1947. For them, the struggle of the Democratic Party, the Sereikar movement, and the Khmer Republic represent one continuous effort to establish democracy in Cambodia. Other intellectuals came from the political elite under Sihanouk and the *Sangkum* that was toppled from power by the democratic intellectuals and politicians of the Khmer Republic in 1970. As a result, the royalist intellectuals could be found in the Sihanoukist Funcinpec and its military wing ANS. For the most part, these pre-genocide loyalties were decided over their membership in either of the movements – with only few exceptions.

Habitus Formation and Lifecourse

There were roughly two types of intellectuals who joined KPNLF and Funcinpec. The first type were highly educated, well situated, and mostly also politically well-connected civil servants from ministries or politicians and activists. They acquired university degrees from one of the very few Cambodian universities during the 1940s or 1950s, or even from the US or France, in medicine, engineering, or other mostly technical studies. Basically, depending on their age, they became either civil servants under Sihanouk and then Funcinpec members, or they were civil servants under Lon Nol's Khmer Republic and then members in the KPNLF. They belonged to an old political elite in either the Sihanoukian government or the Democratic opposition, i.e., Lon Nol. Most of them had high volumes of inherited cultural and accompanying economic and social resources that largely prescribed their lifecourse and position within society. The second type were younger graduates from universities during the 1960s and 1970s, most of whom were drafted to the military as high-ranking commanders under Lon Nol to fight the Khmer Rouge between 1970 and 1975 or were forced to stop studying when the Khmer Rouge took power. They were slightly less well connected but still highly educated and politically affiliated due to their family background, with parents and/or grandparents serving Sihanouk in the royal palace or serving the Democratic Party and opposition, respectively. Others managed to acquire a considerable degree of cultural resources because they hailed from an upper stratum of the peasantry (those who could afford to send a child to school) due to scholarships they received for higher education. Intellectuals from both groups emphasized their higher education, the importance of education in their upbringing, and the fact that they did not join the movement but were 'asked' to become a member, a commander, or an instructor for the resistance. Thereby older patrimonial elite networks and age-old divisions between them were reconstituted. While the first group's status rested upon a large volume of inherited resources, the second's inherited resources were a bit lower at the beginning. They acquired much of it due to the educational reforms and their capability to obtain scholarships.

Intellectuals of the first type were oftentimes sons of famous and long-serving opposition leaders or government officials, coming from a handful of leading families ruling the Cambodian elite or families and individuals closely associated with these circles. Even before King Sihanouk's massive widening of the educational sector during the 1950s and 1960s, they graduated from one of the few high schools (mostly attended only by elite children) and either continued on at one of the even fewer universities in Cambodia or got a scholarship to study in France or the US. With a university degree in their pockets and their family background, they received a more or less predetermined position in the Cambodian society at ministries, as high-ranking civil servants or leading party positions. Most of them commuted between Cambodia and France or the US. Some even stayed abroad for several years in Diaspora communities, having a second home abroad. After the Khmer Rouge took power, Cambodians staying abroad at their 'second home' or for study purposes suddenly became stateless and, in turn, refugees (GS-KP1). While the Khmer Diaspora in France was slightly more attached to the royalty, the Khmer Diaspora in the US was (and still is) closer to the Democratic movement.

Commuting between Cambodia and various countries abroad was not an option for the second group of intellectuals, who had contacts to leading figures in the elite and came from well-off but rural business, teacher, or mid-range civil-servant families. After the opening of many schools and universities by Sihanouk's educational reforms, they and others, including some children of the elite, gained access to prestigious schools. Their families struggled to finance their studies, and many had to earn an additional income even though they were comparatively well off, thereby coming into contact with influential people during their studies. But being young and still students or recent graduates meant that many of them did not have the position and resources to flee when the Khmer Rouge took power. While the first group of intellectuals largely went abroad to places they had been living in or commuting to for many years or had family overseas, the younger intellectuals stayed behind and struggled to survive the years of genocide as 'new people' or 'capitalists' on the Cambodian rice fields and in construction projects throughout the country. And while former soldiers had almost no choice but to flee, these intellectuals tried to hide their background and to survive the regime by working hard for the Communist organization, or 'Angkar'. Some others were simply 'lucky' enough to be staying abroad on a scholarship to pursue their studies during the takeover by the Khmer Rouge. All intellectuals, however, stressed that they were, as one said, "above people, not equal to them" (BC-F3), and those who stayed behind now had to "live among the ordinary" in order to survive 'Democratic Kampuchea' (BC-KP2).

Access to and Mobility in the Field

While the Diaspora intellectuals largely created or joined the KPNLF, the intellectuals from the interior who survived the Khmer Rouge joined both non-Communist organizations. However, their mode of access was different. The Diaspora intellectuals stressed the choices they had after being 'asked' by the leadership to join the military and completing the Cadet School in Beoung Ampil camp: "You can (1) serve the infantry (.) you can serve the guns unit, you can serve the operational staff, you can serve (2) you have the CHOICE (.) you can serve in the general staff, which means running the operation from the office (.) or you can chose to be a field soldier (.) joining unit, yeah (2) so it's up to you". While they had the choice, he stresses that it was a "logical choice" for "the refugee" to become a rank-and-file soldier (RC-KP1). Many changed their position almost on a yearly basis, being in charge of different offices and units and having a strong desire to lead special military units for psychological warfare and intelligence. Their contacts opened every door along the border, even literally: "I traveled to KPNLF zones only with key personalities" (GS-KP1). The intellectuals from the interior, however, accessed the resistance one echelon below. They knew some leaders but not the highest ranking ones. And many first hoped for a change by the Vietnamese, whom they greeted as liberators at the beginning. Some even went to Vietnam to receive political instruction. But then they saw that there would be yet another Communist regime, only this time controlled by Vietnamese cadres.

Hence, many took off to resettle in a third country. Fleeing to the border, most did not really plan to join the resistance but thought to get to the US or France as refugees, which is why recruitment of 'intellectuals' among the 'new people' was not very easy. Most were fed up with starvation and fighting, seeing no real end to the long civil war that lasted decades. Others, however, heard about the resistance and knew some people there, basically from the schools they had attended between 1970 and 1975 and whom they regarded as their 'sponsors': "I knew a lot of people. I had friends and godparents all over Thailand. I was not afraid of having no rice to eat there. I knew a lot of people" (RC-F1). These godparents introduced them to military commanders, who in turn made them mid-range commanders in battalions or companies. They joined 'by networking', as knowing an influential person meant that you could procure a position one rank below that 'sponsor'. These intellectuals did not have any military background prior to joining the CGDK. But now they were asked to join the Cadet School for basic training and become high-ranking brigade or regimental commanders. Recruitment by network worked much better and introduced some intellectuals from the interior to the resistance, while those who were not aligned with someone within the patrimonial network for the most part simply sought a quick resettlement. Old patrimonial elites reconstituted their networks to get the country and their leadership position within Cambodian society back.

Using Pierre Bourdieu's scheme, the intellectuals' main field resource would be a cultural resource: academic education. Hence, education or 'intellectual superiority' is the resource structuring the intellectuals' discourse and practice. However, within the field of resistance, they now had to defend their high status against military commanders and strongmen from lower social milieus. Already their earlier lifecourse had been characterized by a struggle to defend their privileges and status, after being pulled out of the country by the Khmer Rouge and deprived of their citizenship while staying abroad. But their habitus was still characterized by a belief in a 'natural', inherited, and thereby secure difference that makes a difference to the rest of society, and other commanders in the field in particular. This difference put them 'above the ordinary', despite of their struggle to maintain their leading status during their lifecourse. The main resource they used to defend their either inherited or acquired status was their high degree of education. While the value of their main resource was questioned by the strongmen, for instance, it was not field-bound and, therefore, was by far not as volatile as was having fought many battles.

Classification: The Body and the Brain

Respondents from this group all classified themselves as "intellectuals", having studied subjects like medicine or engineering and pointing out that they were "above people, not equal to them" (BC-F3). Most of these individuals, of course, studied abroad and picked up concepts of democracy, the rule of law, or management techniques, and were fluent in English and/or French. All drew a very clear elitist line between themselves and "the ordinary people", i.e., the "mass of refugees" along the border (RC-KP1). While these commanders were intellectuals, "a typical resistance soldier was a young single man with little or no education, who has no commitment at the moment to any other things in life" (GS-KP1). The intellectual's classificatory discourse is put in a nutshell in a sequence given by a regimental commander from the US:

> During my five years of military soldier I never killed anyone (1) even though I was commander of the operation (.) I got a pistol in my pocket (.) but never used it (.) I used it just to exercise (.) to shoot (.) you know (.) just in case. But I had my bodyguard (.) and when there was a bomb exploded (.) my bodyguard are on the foot (.) READY TO ENGAGE (1) but I had to STOP them (1) I said (.) try to find out, where does it come from, who shoot that, how far is it from and where are other elements and so on and son on [I: The training comes back] YEAH (1) so I approach in a very LOGICAL sense of way you know. But then (.) NO (.)

you see THEM (.) JOY (1) whenever there is a fighting, they are ready to fight (.) but asking questions like fight for who, fight for what, fight where and this and that and so on (.) it is just the mindset (.) they turn to that. (RC-KP1)

In short, one might interpret this to mean: The soldier is the body and the commander is the brain. Speaking further about the mindsets of rank-and-file soldiers, he adds, exemplarily for the group of intellectuals, "without any pride at all" that "NORMAL Cambodian people are skillful fighters": "I mean they REALLY love gun" (RC-KP1). In addition to being ready-made warriors craving battles but substantially lacking logical-intellectual guidance, a Funcinpec commander highlighted the rather convenient aspect that they are also obedient by nature: "[Obedience] happens automatically. Soldiers always listened to their commanders" (RC-F1). When asked what a commander had to do to make his soldiers follow loyally, it is rather a surprising question for an intellectual: "I did nothing. I was appointed. I was a commander" (RC-F1). Luckily, the loyal automatons at the rank-and-file level have good commanders giving intellectual and political guidance about what they are fighting for and how, never having any "doubt that the peace is really at the end of the tunnel" (RC-KP1). The commander's position in the field is almost given by nature, his superiority lies beyond any volatility and there is no need to engage in disciplinary practice. There is no need to transform recruits into loyal and skillful soldiers or prepare them for battles of any sort; since they are not afraid, they even enjoy fighting and are warriors simply by being "normal" Cambodians. However, some also highlighted that it is not very difficult to be a soldier: "As long as you knew how to pull the trigger, you could join the combat".

There is only one case of indiscipline imaginable for the intellectuals: hunger. Hunger and related physical needs serve as explanation for those few cases of indiscipline that occurred from time to time. Asked about the source of indiscipline of "that few" and how to deal with it, a typical answer was: "We had to make sure soldiers would not get hungry" (CL-F1). Hunger made people act desperately and rob or even kill other people. Sometimes it was also some trouble with the family that made them lack discipline: "The first reason was because of stomach problems [hunger] and second, because of trouble in the family" (RC-F1). Furthermore, the intellectuals are also special because they have special knowledge, which an ordinary soldier lacks, especially regarding the dangers the Vietnamese (or, using a common derogatory term, the 'Youn')[3] pose to the country being swallowed. Studying Cambodian history taught them: "We were thinking that our country was full of Youn and that Youn had been taking our territory throughout history since King Jayajetha the Second and the period of Kampuchea Krom" (BC-KP3). The discourse on intellectual guidance and superior knowledge that is needed to make reasonable use of the physical force of the rank and file also serves as a foil to mark a difference to the strongmen and their 'warlord' leadership style.

---

[3]   A regiment commander from the Khmer Rouge had his very own theory what the term means: "When they [the Vietnamese] invaded Champa and did not leave the country again, the French named them Youn, which means 'earthworm' in Tonkin Chinese ((laughs))" (RC-KR1).

Field Conflicts: Militarism, Warlordism and 'Old People'
The intellectuals faced skepticism when joining the military side of the resistance, being addressed as 'arm-chair heroes' (Heder 1980, 87) and being under scrutiny over their goals as "ex-pats": "For an ex-pat like me, there was an inevitable period of close scrutiny as longer-term members judged the newcomer's commitment to the organization and its goals" (GS-KP1). The militaristic atmosphere of 'discipline' and 'toughness' did not provide much space for 'creativity' on the intellectual's side, as one intellectual outlines his first thoughts when joining:

> I thought, 'Pathetic'. I worried whether any change could be made. 'Keep your mouth shut', I was reminded time and again. 'Loyalty' and 'discipline' superseded creativity and innovation. I was conscious that a deviation would raise question of trust, hence, a return to square one' [...] You can't really see progress when you sleep on the floor with your head on a smelly pillow (passed from person to person) and half of your body under your desk, sharing a room with up to 7–8 people at a time, sharing communal meals, and declining offers from friendly supporters to install air conditioning unit to ease the heat. And you have to develop a very thick skin to smile at jokes and insults in the field aimed at making you 'tough', or so they said. (GS-KP1)

Being a warrior remained strange for these people. They continued to look for opportunities to do something more creative, political, and intellectual. Formalism and ranks went against their very own understanding of progress, but in the end, they felt like they had to adapt:

> Anecdote: I objected to military ranks in the volunteer guerrilla army – I set an example by being the only high ranking figure in the Front's military who declined stars and bars. But when other commanders' men received their promotions as officers and my men felt short changed, I closed my eyes and bit my tongue. I let my men receive their stripes, too. Until I left the Front, I never accepted a rank. (GS-KP1)

While the formalistic and disciplinary space of the military was too narrow for the intellectuals, the strongmen and those subordinates who were potentially categorized as 'old people' under the former Communist regime posed yet another problem. For the intellectuals, the strongmen were uneducated and brutish warriors who could not be trusted with the political and strategic guidance of the movement. Especially soldiers who came from an illiterate poor farmer's background were met with skepticism. Former 'new people', or intellectuals living under the Khmer Rouge and who were still traumatized by their experiences, even feared people who might have been classified as 'old people' and whom they identified as main perpetrators during the genocide. Many intellectuals saw in them numb and wild killers, responsible for mass executions (cf. Heder, 1980, p. 89).

Some intellectuals, therefore, dreamed of educated soldiers, capable of a very different kind of warfare – warfare on a politically and psychologically conscious level:

> I had a different concept of a resistance soldier: He is a guerrilla fighter who should have basic knowledge of the military (his weapons, his environment, the bat[t]lefield), as well as basic capacity to use logic and some psychology in warfare. I wanted a 'good and skillful' guerrilla to be more of a well-rounded soldier. Of course, he should at the least have a primary school education if not a secondary high school education. I was eager to have guerillas with a level of literacy that allowed them to employ logic and reasoning. In fact, the Armed Political Propaganda and Clandestine Operations [...] lacked those elements – a weakness almost impossible to remedy. I knew since my first day in the resistance that I was going against all odds to dare to dream of 'transforming twigs into metal', to borrow a friend's advice. (GS-KP1)

For this different kind of soldier, educated and capable of logical reasoning, "[b]asic military training is a must; political-psychological training is a plus [...] Ideally, I would like my soldier to be literate and be capable of simple logic and reasoning – something hard to achieve since teaching men to think and reason is a long-term endeavor" (GS-KP1).

Guidance, Management, and Psychology of War

> I was more a theoretician, a teacher, a thinker than a commander leading men in battle. (GS-KP1)

For the intellectuals inside the KPNLAF and the ANS, the war was not about battles in the field but, rather, about battles in the minds of the people. Unlike the strongmen's approach, the intellectual leadership highlighted the importance of war as being fought on a psychological level and something that had to be "administrated", "processed", and "managed" well, no matter which kind of informational flows were to be regulated and directed properly. A regimental commander of the KPNLAF, for example, describes the functions of the five bureaus of the army as a question of managing informational flows:

> Managing people, managing information, managing and implementing combat operation (1) study the enemy, study yourself, and how to prepare the operation (1) managing logistics (.) whatever it is (.) ammunition, clothes, tanks, drugs ((laughs)) (.) whatever (.) money (1) and managing information in propaganda (.) the fifth function. (RC-KP1)

Hence, governing a guerrilla force was all about a management of flows of people, information, and goods. Planning became central for leadership and for success in battle: "Good planning, good organization, good leadership, good logistics, and trained soldiers" were imperative (GS-KP1). Good planning rests upon "good intelligence and the cycle of planning, collecting, processing, and comparing data" (I-KP1). Knowledge, management, and data collection are the keys to success in warfare, not bravery. Many intellectuals therefore referred to the teachings on warfare by Sun Tzu:

> You know because according to Sun Tzu (.) know the enemy and know the self (.) that battle hundred win ((laughs)) if you know only the enemy (.) you win but (.) win for failure in the future (1) (I-KP1)

Studying the enemy and your own forces closely is a prerequisite of any attack and military operation:

> Our saying says 'Know yourself and know the others'. We had to clearly know our enemy. We had to know their location, how many guns would be used, and from where they were supported. If we wanted to attack them, we had to know how we ourselves would be supported. We had to know this because this was provided in our training. This was called 'location attack training'. For ambush attacks, we had to know where our enemy liked to go, and how they walked. (BC-F3)

Management here refers to the allocation and preparation of goods but not to the mental preparation of soldiers for their operation. Preparing them was easy and more or less redundant since they knew what to do by themselves: "It was not hard to prepare them. They knew it all by themselves. When a soldier carried a gun, they knew that they had to approach their enemies. Then they simply shot or did whatever to bother the enemy", and by defending themselves, the warriors' competence increased automatically and the leader merely had to collect data:

> They [the soldiers] could not allow themselves to be shot, so their competence was growing automatically. Before going for combat, we only had to know how many soldiers our enemy had, where they have been stationed, which guns they had, (1) which bullets they used, how many soldiers have been deployed. (RC-F1)

The character of the soldiers was beyond organizational control for the intellectuals. They did allow them to be part of the non-Communist resistance, but many aspects of their behavior were simply part of their nature, impossible to prevent. Rape, for example, is part of 'human nature', as a Funcinpec brigade commander outlined by comparison to American soldiers: "This is human nature. All soldiers, even American soldiers, who are known to be professional soldiers, raped Japanese and Afghan girls. We could not stop it as it is human nature" (BC-F3). The KPNLF's 'ocean policy', for example, meant that the leadership let everyone in, but far from being responsible for their actions, the leadership saw itself as a good force "clearing" the ocean but not responsible for the character of each and every soldier:

> The ocean receives water from large and small sources, [...] be them clear, muddy, stinky, fragrant, dirty, [...]. But the ocean has one main duty: to make the water clear and salty. Likewise, the KPNLF receives all types of people, be it scholars, intellectuals, students, rich, poor, ignorant, educated, workers, farmers, religious men, monks, nuns, criminals, thieves, robbers, nurses, teachers [...]. They are all welcome. (Kong 2009, 195; English translation slightly corrected by the author)

Good management therefore lies in the limits set by the leadership and the correct composition of forces, not its ability to change people to become better soldiers (which is simply impossible). It is the correct distribution and use of given entities and forces:

> There is our Khmer saying that 'The curved object makes bicycles, while the straight object makes spokes'. The importance lied in the fact that commanders had to know which skills and capabilities they had. Leadership was decisive since there were all kinds of rank-and-file soldiers, be it bad or brutal. They all lived together in one group. But when the commander was clear, there was a limit to what the soldiers could do. (BC-F3)

Leadership does not transform people but makes correct use of their nature and the given and ready-made realities they provide. The ready-made warrior has some good or convenient characteristics (such as being brave and obedient) but also has some negative characteristics (such as a tendency to rape civilians). Hence, he needs the guidance of an intellectual who knows how to arrange and direct the flows of goods and people, and their characteristics, thereby creating something good and clear. The discourse on the ready-made warrior being "born brave" (BC-F3) inside the intellectual's habitus group is not only a vivid example for a difference made to legitimate a difference in the social and field but also leads easily into prevailing power practices used by the group to control their soldiers. In doing so, the discourse frames two main types of power practice.

Power Practices and Institutions

Intellectuals acted from a greater distance from their troops than the strongmen, maintaining a clear physical separation to their own subordinates. Soldiers knew them only by hearsay or met them by accident, which is why most could not say much about their boss: "I: What was [name] like? R: I don't know because he was a high-ranking official. I rarely met him. I only knew that he was our commander". He only met his commander once in a queue for receiving food but was not permitted to talk to him: "Commanders discussed among themselves, and our job was to be ready for combat operations, get our materials, and jump on the car" (GC-KP2). Acting from a clear distance, the intellectuals' discourse translated into two main power types: incentive practice and security. What is striking about the intellectuals' practice, however, is also what they did not do. Basically, they skipped every classical disciplinary technique and preparatory step for combat. Stemming from their secure field position, they felt no need to ensure obedience within their ready-made and combat-craving followership.

Disciplinary Techniques: Skipping Training and Combat Preparation
Asked about trainings for rank-and-file soldiers, most intellectuals answered that it was more or less 'learning by doing' for those being recruited "from the refugee" (RC-KP1).

> And as for the training (.) not much of training, you know. They do (.) they experience [hm] (.) they go out and shoot at the Vietnamese ((laughs)) and thing

like that. (1) And they PRACTICE (.) by DOING practice. There were some training I think. But it is only leadership (.) that I know. (E-US1)

For a rank-and-file warrior, there is not much that he needs to be ready for combat:

> He needs basic training – and more training; he needs weapons and material; he needs to be sure that his family left behind is in good hands – or he may be going into combat operations leaving his head and his heart behind. He should be physically, materially, and morally equipped. (GS-KP1)

This basic training that was provided in the KPNLAF after a couple of years operating at the border consisted mainly of physical exercises: "We only had to make sure that they are physically strong" (BC-F3). Therefore, the soldiers received some collective instructions on how to crawl, to salute, to hide, and to shoot using a wood stick as weapon. While the KPNLAF had a Cadet School, in which commanders from the company level onwards were taught leadership skills and strategy, rank-and-file soldiers, were, at least theoretically, trained in a special school, in which they went physical training in collectives. Funcinpec sent most of its commanders abroad for training or to the Cadet School as well. Only later on did the general staff of the ANS have its own training center for high-ranking and mid-range commanders, as well as an "Instruction Center" for light physical exercises for the rank-and-file soldiers (BC-KP3). Training ready-made soldiers to fight seemed to be useless anyway. At best, it could be compared with teaching letters to someone who can talk already: "This stuff did not have to be trained. Everyone tried his best to succeed. It is like teaching consonants to students. We already know how to spell and read vowels and consonants" (RC-F1). Hence, such trainings were seen as merely complementary practice.

While military drill was not of interest, the intellectuals did set up political indoctrination courses and institutions. Political-intellectual guidance was provided by regular "conferences and discussions" (RC-KP1). The KPNLF even had a Political Training School with instruction on moral and political matters, such as not to spoil civilians, not to rape their daughters, or related advice: "Nothing too complicated was taught, just that" (SC-KP1). Besides being instructed on morality, the rank-and-file soldiers also learned about the political view and vision of the leadership, what they were fighting for, against whom, or about how the Vietnamese (or "Youn" as they were constantly called) constantly tried to swallow Cambodia. Or as a Funcinpec commander said: "You know, as in any school – indoctrination why they fight" (BC-F1). The leadership was in a position of knowledge about the secret plan of the 'Youn' to erase the Khmer race, which they had to share among their subordinates: "I am speaking frankly here. The Youn were everywhere in the country. We had studied history, so we knew" (BC-KP3). Since warriors do not care what they are fighting for and live in and for the moment, the intellectual leadership had to take care of the wider and greater vision. They live a life without hope and tomorrow (as contrasted to Thai soldiers, for example, who had a home and a family): "And the Thai soldiers fear Cambodian guerrilla soldiers because of the guerrilla (.) don't live their life with hope. They EXPECT to be killed (.) today (.) tomorrow. They don't have future (1) they don't have tomorrow" (RC-KP1). Wartime psychology makes them shortsighted and even more in need of political indoctrination about the major cause for which they are risking their lives.

Incentive Practice

Since tomorrow is not a given, it does not make much sense to promote a rank-and-file soldier. By contrast, it is much more reasonable in this framework to give them something to enjoy in the moment. While rank promotion for rank-and-file soldiers was rare, the main instrument that intellectuals employed to make soldiers obedient was to feed them well and to give some incentives, thereby covering their basic needs to prevent desperation and hunger:

> Whereas many in the Coalition Government of Democratic Kampuchea – especially the Khmer Rouge – put new civilian and military recruits through military training and political indoctrination to make them loyal fighters for the resistance, I found myself a loner here. Because there is no salary in the resistance, the best compensation is good treatment, and more good treatment, of a soldier and of his family, with compassion, empathy, respect – the more so as he may be an illiterate peasant who agrees to put on his uniform and carry a weapon for your cause. Make sure he and his family have the basic necessities to live – a thatched roof, food (rice, dried fish, a pot), clothing (even used clothes), medicine (a dispensary), and, God-willing, a little cash ... Then he is likely to owe his loyalty to the cause. (GS-KP1)

Hence, he goes on: "Better and better treatment is the best reward. And more financial and material rewards to a soldier and his family are very helpful – even more than his promotion" (GS-KP1). Most intellectual commanders talk in length about providing the basic needs to soldiers as sufficient means for establishing and maintaining discipline – although that was not really difficult due to the kind support by the UN:

> Everybody give money at that time (.) yeah. And also the commander give more food, more money (.) and also to the wife, the children, who live together in the camp (.) you know. So by giving good to the camp (.) to the family (.) you also strengthening the morale of the soldiers (1) for more fighting you know. (2) But most of them (.) they are well fed, they have enough food. (1) They have clothing, they have medical care you know (.) from the United Nation. (BC-F1)

Focusing solely on situational means, the rewards to enjoy for a certain success were extra food, extra money (no salary), or some time off of duty to visit the family. The high command of the KPNLF even established a cash system after 1985 in order to maintain the morale during the leadership quarrel, in which a soldier received a fixed amount of money for shooting a tank, hijacking a bridge, or for killing or arresting some enemies. Funcinpec furthermore distributed letters of appreciation by Norodom Sihanouk. Also, from mid-1980s onwards, mission packages were distributed to encourage units to actually go for a mission, which, by contrast, in some cases may have been the reason that some units stayed at the back and pretended to have done some fighting:

> We [company commanders] were given three thousand baht each. The [battalion] commander got twenty thousand baht. Fifteen of which would be used to buy food, and the remaining five were for the commander. (CC-F1)

All rewards, however, remained situational and individual. There were no collective rewards such as those used by the Khmer Rouge leadership, the strongmen, or the old military elite. The intellectuals kept close only those who were educated like themselves. Patronage as a dyadically secured exchange existed only as career sponsorship between 'the educated' on a higher level, not with the 'illiterate' rank-and-file soldiers, thereby closing the two social spheres. Moreover, although the intellectual commander 'took care' of the needs of his subordinates as well, and therefore 'patronized' them to a certain extent, he really just tended to the basic, physical, and rather urgent needs.

Punishments
Talking about punishments is difficult for the intellectuals, who constantly work for and propagate the value of human rights. Therefore, some stressed that they preferred to set misbehaving soldiers free to go to the interior, to do something else that was more suitable for them:

> Because of time and material constraints in resistance warfare to try to correct misbehavior, I believe in doing what is possible to encourage the misbehaving guerrilla to exit the resistance and help him find or start a new life (outside of the resistance) doing something else – go be a fisherman, a woodsman, a farmer – or go for a resettlement elsewhere. (GS-KP1)

Moreover, when confronted with wrongdoings, some preferred threats rather than real hard punishments: "We had to threaten them that they would be killed if they raped a civilian daughter. However, the rule was taken very seriously" (RC-F1). Oftentimes, misconduct could go unpunished.

Punishments, however, touched the main question of obedience and control and caused major disruptions between the intellectuals and the strongmen in particular. While the strongmen used seemingly arbitrary violence to instill discipline, following only their very own judgments and goodwill, intellectual commanders and the higher command under Son Sann constantly tried to implement a formalized 'rule of law' and a judicial system. For example, Son Sann ruled to implement so-called "Justice Committees" with institutionalized disciplinary procedures and fixed prison terms for each kind of wrongdoing (cf. Lawyers Committee 1990, 140–47). By formalizing the punitive procedure, mistakes were supposed to be ranked according to the severity of the offense: "We punished them according to the degree of the mistake they have made" (BC-F3). However, not only did most of the strongmen and also those from the old military elite ignore the advice on how to create such committees, complaining that doing so would make the units open for 'external criticism', but also the advice in itself remained largely abstract. Misconduct of higher-ranking commanders went largely unpunished due to the powerless position of the leadership in front of the 'warlords', as a regimental commander admitted after considerable contemplation:

> I: Were there any punishments for misconduct?
>
> R: There were (6) yes (1) there were but (5) but again (.) we were in a wartime (.) you know (2) if you got caught raping people (.) you know (1) if you are a NORMAL soldier (1) rank and file (2) you are arrested (.) put in jail and so on (2) if you repeated many time (.) you could be killed and so on (2) but if you are important person like chief of company (.) you know (.) commander of company, battalion commander and so on (.) the command structure tend to close their eyes (1) and discretely tell them not to repeat again or move them to another camp and this and that and so on. (RC-KP1)

In the end, the right to punish remained largely in the hands of regional commanders and camp leaders (CC-KP1). At times, all that was left was to pretend that someone was punished – at least in front of the civilians demanding justice:

> When people came and complained about something [the soldiers] did, we pretended that we were really mad with them. We ordered other soldiers to arrest those thieves for execution. Then, the villagers felt calm again. They thought the thieves who had stolen their cows would be punished. We had to act that way as if we were really serious in front of the people, while in reality, we were not. (RC-F1)

While the implementation of a 'rule of law' largely failed, the internal discipline of military units was not a problem at all for intellectual commanders. Establishing command was not necessary with warriors being obedient by nature. The intellectual's sovereignty therefore was never at stake and was not reinforced by punishments as it was with the strongmen. In contrast to the strongmen, therefore, intellectuals did not punish severely if someone refused their orders: "For example, in case they didn't follow orders because they were too afraid, we would analyze the situation. We did not always punish disobedient soldiers" (BC-KP3). The only case when punishment became harsh and was taken very seriously was when the reputation of the movement as such was at stake with civilians and other people outside the resistance. Then the style of punishment became highly physical (sitting in the sun for hours, beatings, sometimes but very rarely killings) and not mentally re-educative. Among the highest-ranking crimes was raping a *civilian's* wife or daughter (not raping a *woman*!), which also stands in stark contrast to the practice of the strongmen, who simply 'asked them to stop':

> Most importantly, they were not allowed to threaten or rape other people's wife and daughters. They had to be morally good, and we also imposed punishments on them for this. If they raped others, they would be sentenced according to rule

of law. The importance lay in the political dimension that soldiers had to be nice to the people in order to gain their support. (RC-F1)

Punishments and the 'rule of law' in such cases had a central function: managing the effects of soldiers on civilians, not securing obedience among the followership. It was basically to manage the impact of collectivities on each other (resistance – civilians) regarding the political goals of the movement. This leads nicely to the second power type as practiced by the intellectual commanders, namely security.

Security: Managing Flows and Their Effects
According to Michel Foucault, the type of power that he calls 'security' manages multiplicities as a series of given, unchangeable ("ready-made") elements. It "works on the given", thereby maximizing positive effects and minimizing "the risky and inconvenient" effects (Foucault, 2009, p. 34). Instead of subjugating individuals like in sovereign power, security practices attempt to govern collectives of any sort, such as goods, people, or information, by regulating their modes of circulation. Power thereby becomes not just a negative force of subjugation but a positive force in creating collective states and regulating flows. Approaches analyzing power solely as a negative force would omit this technique. Security chooses measures by relating its effects to collectivities and collects data in order to keep track of its states and developments within them.

In the end, security is the core of the intellectuals' power practice. While the ordinary rank-and-file soldier was a given, brave, and obedient entity that did not need much attention and preparation, the regulation of informational flows and their security in (combat) operations in particular had to be meticulously planned, managed, and organized. Operations were only launched after all elements had been at least double-checked: logistics, intelligence, battle supplies, geographic information, and "human organization" (BC-KP4). Every element "had to be known clearly" and prepared well in relation to each other (BC-F3). While the Khmer Rouge forced their units to fulfill a stringent quota, for example by issuing an order to fight three times a day, the KPNLF sometimes took weeks if not months to send a unit to fight. Thereby the intellectuals set up a huge bureaucratic apparatus, given the size of the guerrilla forces in total, leaving many subordinates puzzled about the actual benefit of the paperwork introduced by these 'literates and professors':

> When the literates and professors joined us, they thought they were of high value. So everyone in the resistance started to do things academically. We started to have proper papers and documentations as references each time we did something. We submitted request forms when we wanted one hundred shirts, three hundred hats, guns, bullets, and so on. (RC-F1)

Although the reports did not work out smoothly, the intellectuals set up multiple offices for internal and external intelligence within the Second Bureau in the KPNLAF and ANS. For the intellectuals, the success as a guerrilla faction was decided by an "investigation war", in which the accurate preparation of forces and information was the key: "[Investigation war] was most important, as it was the source of information. No matter whether it was an ideological or military war, its success would only be achieved when we obtained clear, accurate, and well-defined information" (CC-KP1). The Second Bureau was in charge of collecting all the data and maintaining statistics on the performance of military units (defection rates, losses during battle, state and numbers of weaponry, for example). However, most internal material was on "enemy infiltration" (BC-F3, I-KP1). The main responsibility of the Second Bureau – using Sun Tzu's terminology – was to 'study the enemy', instead of 'studying yourself'. The KPNLF had two intelligence research institutions called 'Khmer Intelligence and Security Agency' (KISA) and the 'Bureau of Intelligence Research and Documentation' (BIRD), which was formed later on. The main interest of those documentation and intelligence offices was 'security', be it on the personnel or the documentary level: "The security in collection (.) in collecting uhm (.) intelligence (1) you MUST have security (1) and so I was translate what we call PERSONNEL security (.) uhm and (1) document security (.) you know" (I-KP1).

The main martial goal for the intellectuals, however, remained the battle for the hearts and minds of the people. All intellectuals pinned their hopes on possible elections at the end of the war, preparing and convincing the minds of the people to choose the correct box during a vote. Psychological and political warfare, therefore, lay at the center of their daily efforts, a row of martial techniques that were even taught a specialized school inside the KPNLF: the 'Political Warfare School'. Moreover, within the 'Department of Planning and Analysis' (PLANA) that was in charge of that school and had an office for 'Psychological Operations' in Site II, two programs staffed with elite soldiers were created. Within this department, two special operations were created, called "Armed Political-Psychological-Clandestine Operations" (APPCO) and "Operation SSD". These two programs for psychological warfare were the main pillars for the intellectuals to win the war, and not actual battles in the field. Instead of brutish and brave soldiers, these units gathered the educated and "higher calibers": "APPCO students are recruited from among the 'better educated' civilians at Site II in general, and in some cases, from among the 'higher caliber' soldiers/NCOs/officers with known skills". It specializes in "the higher political indoctrination and the higher level of political consciousness of the KPNLAF civilian cadres and combatants" and it shall fulfill highly specialized and mobile and clandestine tasks with small teams (KPNLAF, 1987a; slightly corrected).

The second program within the PLANA Department of the KPNLAF Second Bureau was called 'Operation SSD'. Missions of these clandestine specialized operations were to establish clandestine networks within the incumbent government, to sabotage physical targets in the interior, and to demolish the enemy's willingness to fight (KPNLAF, 1987b). By setting up these two units, the intellectuals effectively and quite literally separated themselves from the 'ordinary refugees' and directly dealt with managing the effects that the resistance had on the population as well as political and mental preparations for possible future elections.

Old Military Elite

[Being a soldier] has been my calling since I was born. (RC-KP2)

Coming from families with far-reaching networks in the Cambodian military elite, most of the respondents in this habitus group look back on several decades of military service, oftentimes even reaching back to the *Sangkum Reastr Niyum* under Sihanouk. Others are sons of the country's inner military elite circle. However, many of these commanders have been generals and lieutenants with several stripes and fought against the Communists almost since the beginnings of Communist uprisings during the 1960s and also held high-ranking military positions when the Khmer Republic fell to the Khmer Rouge in 1975. General Dien Del, for example, together with General Sak Sutsakhan, one of the main military figures inside the KPNLAF, was one of the last to leave the capital, being pushed out by 'Angkar' on April 17, 1975. Many evaded execution by taking refuge in France and the US, as did leading intellectuals and politicians by that point in time. Others remained active along the border, leading small regiments, or fled Khmer Rouge purges after some months in rather small groups like the strongmen did. All of these commanders, however, were born into the office of a high-ranking commander rather than a rank-and-file soldier due to their family background, making their leadership position up to their engagement in the KPNLAF or ANS largely inherited. They perceived themselves to be born commanders not just due to fate but also simply due to their social position, laying down a highly preset lifecourse.

Habitus Formation and Lifecourse

All of the respondents were the sons of men who served as generals under the king or even before. Many, if not all, even mention that their grandfathers served as high-ranking military commanders as well. Therefore, all of them attended an elite military school. Finishing their Bac I or Bac II degree was not relevant to their success, and some never did complete their schooling. Schooling was not really relevant for their own status; more important were the stripes earned during their service:

> I was an old soldier, and I had knowledge about military affairs as well. Most of us had been soldiers since the King Grandfather's time [Norodom Suramarit, who was king from 1955 to 1960], and we held five, four, three, two, or one stripes until we became Lieutenant General, such as Mr. Dien Del and Chuor Kim Meng. (BC-KP4)[4]

Many from the very top echelon of the military elite's inner circle, however, went to the prestigious Lycée Sisowath during the 1940s and 1950s, at a time when

---

[4]   To avoid misunderstandings: this statement refers to their time within the KPNLAF. Dien Del never was promoted beyond brigadier general as of 1975, and Chuor Kim Meng was a colonel as of 1975.

only children of the political and military elite could attend such schools. Here, the biographies of the military and political elite already intersected. Later on, they served as mid-range commanders of the colonial Cambodian Army and rose in rank while attending several military schools either in Cambodia, such as the Royal Military Command and Staff College and the Royal Cambodian Officer's Academy, or in France at institutions such as the General Staff School in Paris (see 3671302006; 3671302007; 3671302008; 3671302009).[5] The military elite was supported and trained by the French military over decades, and it was only during the Khmer Republic that the US increasingly stepped in to take over military assistance after the French failed during the First Indochina War and the upcoming US disaster in the Second Indochina War. Therefore, most commanders were fluent in French and were highly connected in the Khmer Diaspora in France but had good contacts in the US as well.

When the Khmer Rouge took power, most of the military elite resettled to the US (due to the co-operation with the Khmer Republic) or to France (due to life-long connections within France and the Khmer Diaspora in particular). But life there was very different, and most experienced a downturn in social status when they became 'ordinary' civilians in the US. Others, residing at the border, also preferred to stay on top rather than become a servant for others and maybe even struggle to make a living the US economy: "If I go to the US, I would be working as a subordinate of others. So it would be hard. And I also would have to spend a lot of money on daily expenses" (BC-KP1). Soon the old commanders started to set up Diaspora organizations, forerunners of the resistance, in cooperation with the intellectuals and politicians abroad. After the Vietnamese invasion, these old military commanders had only few choices available to them. Going back to Cambodia with yet another Communist regime in power was not an option, not just because of their own anti-Communist attitude stemming from their life-long struggle against Communist rebels but also because they would not have been warmly welcomed by the new leadership in Phnom Penh either. Leading the resistance as they have been asked by the political side of the Diaspora was a good opportunity to revive their status as the elite of the Cambodian military apparatus and to prepare for their return. All they needed was a military to lead.

Access to and Mobility in the Field: Regrouping of an Inner Circle

It was not primarily their educational level that was decisive in making these commanders leaders of the upper echelon of the resistance but the simple fact that they had a long record with the Cambodian military elite: "Some didn't [have Bac II] as they have been known to be old commanders. Those who used to be soldiers during the King Grandfathers or Lon Nol regime were tolerated [even without the degree]" (BC-KP4). Oftentimes these commanders were accepted simply due to their rank in the old regime, their military knowledge, and their inherited family

---

[5]   The number relates to the database of 'The Vietnam Center and Archive' in the reference list.

background. Some of the youngest never even attended a school before the Khmer Rouge took power, nor did they have any experience in commanding soldiers of any sort. Their status was largely inherited, as they were born sons of a former high-ranking general who had served the government for decades and/or generations. They were "raised in a military barrack", and in some cases that was enough to be part of the inner circle (BC-KP1). By contrast, older commanders in this circle had been organizing the resistance against the Khmer Rouge rebels already during the early 1970s. Due to their age and experience, being expressed in the number of 'stripes' as main symbolic resource and the herewith connected military rank, they occupied slightly higher field positions than the younger members of the elitist group.

Commanders from the old military elite became either members of the non-Communist general's staff, responsible for operational planning, heads and instructors of the military training schools (such as the KPNLF Cadet School at Boeung Ampil), or brigade commanders. Those at the training institutions and the brigade commanders who were directly in charge of disciplining and commanding soldiers are most relevant to this study. However, most of these commanders served both functions either at the same time or subsequently. The old military elite had been entrusted with scripting basic organizational rules and regulations for the two non-Communist militaries, especially inside the training camps. They set the rules from scratch, relying on their own recollections, a few old documents, and information collected across several countries. Everything, including the flag and anthem of the resistance, was made from scratch and sometimes simply based on experiences gathered by spending most of their life in military barracks:

> Why I knew military rules and techniques was also because my father was a soldier, that's why I lived most of my life in a military barrack. Together with other senior officials, I was compiling all these strategies and rules, which have been used until now. I also drew our national anthem and flag. (BC-KP1)

While those commanders with a long personal record of military leadership acquired a position in the general's staff and as commanders-in-chief, it was largely the family heritage as a social resource that decided their position in the field. This reproduced a military elite caste and its loyalties across a tiny inner circle. Only among these commanders did factors such as military education, age, and experience matter and influence additional hierarchies among the inner circle. For the military elite, the main symbolic resource was their military education and its symbolic value as materialized in 'stripes'. The military elite's discourse and practice, therefore, revolved around military expertise, discipline (respect for the chain of command), and drill for the rank-and-file soldiers.

Classification: Drill, Discipline, and Proper Strategic Planning

Commanders from the old military elite have a very classical picture of their own soldiers. For them, a soldier is just like formless clay, or like a blank sheet of paper: "Soldiers were just like blank paper" (BC-KP1). It is military drill and training that forms them into a good soldier. The recruit himself has almost no part to play in this process, except for to indicate the 'temperature' he prefers for his formation, as one commander noted:

> When I was commander back then, I divided our troops into three types – cold, moderate, and hot. For example, we had one thousand soldiers. Three hundred and thirty three preferred 'cold', another three hundred thirty three preferred 'moderate', and the remaining three hundred and thirty three preferred 'hot'. We had to divide them and remember the separation clearly. For some soldiers, we did not have to beg them much to do anything: 'Dear, you should do this and that'. With others, we did not even have to say much. They would do it right away. For some others, however, we had to kick them first before giving commands. Once I was shot in Boeung Ampil. The soldiers who were there to help me first were the ones I had been kicking before. I sometimes hit soldiers in the entire brigade. I asked them to queue up and I hit them all. However, those who I hit came first to save me. I think I did the right thing. I did not kill any soldiers or punish them too severely. I just instructed them. If you don't believe me, you may ask my soldiers at my work place. Each of my soldiers knows that I am vicious. Sometimes I hit them so hard that the stick broke, but I was actually advising them. It doesn't mean that I hate them. After the punishment, everything was fine. I would forget all their mistakes, and start anew. I also supported them. Especially my wife. When my wife saw that I was hitting soldiers, she stopped me from hitting them. This was one. Another one was that when soldiers' wives and children were starving, my wife helped. That's why many people liked me anywhere I went. (BC-KP1)

Physical instruction is most important to this commander, and his wife took care of his soldiers so that they still love him despite of his severe 'instructions'. Most commanders from the military elite divided their soldiers in different types according to their ability to develop firm obedience as rooted in their nature. Those who were bad and misbehaved simply either needed to be given instructions in an aggressive, 'hot' manner or it was assumed that it was impossible to form them into loyal automatons, comparable to the black sheep that can be found within any family:

> That could be compared to a family consisting of five members, four of whom are usually good, and one of them usually drives a motorbike without mirrors, and acts like a gangster by having his ear pierced. [...] That was their own morality and culture, their personality or race. (RC-F2)

Statistically speaking, a certain share is always spoiled and does not properly respond to drill. And that drill had two main functions: to make them respect ranks and to physically prepare them for battle. Similar to the intellectuals, the old military elite saw the lower ranks as mainly a physical force, something that could be drilled by physical punishments and that primarily needed physical fitness to endure battles: "The most important thing was physical strength, health, and flexibility because those who shot first were those who won" (BC-KP4). Discipline in terms of adhering to and respecting military ranks as visualized by stripes had to be followed within all ranks (BC-KP1).

For the old military elite commanders, when they joined the non-Communist forces, these troops first and foremost lacked classical military training and drill. They lacked their very own military and organizational expertise. Although rank-and-file soldiers were essentially brave and honest, they lacked basic military skills: "There were not many skillful fighters, but they all had bravery and honesty. They were not skillful in fighting because they had not yet received proper training" (BC-KP1). When they arrived, they saw the chaotic state of the scattered troops along the border and the strongmen who commanded them:

> At that time, military discipline was declining a bit because we didn't have a proper military school yet. It was not declining in the sense that they killed each other. First of all, the decline was because the soldiers' hair was long and secondly because they were putting on a lot of make-up. And thirdly, it was because they were drinking too much alcohol. Especially the commanders drank a lot. (BC-KP1)

As long as they knew how to obey orders and to implement strategies as laid down by their leaders but did not obey commands, soldiers were not good enough.

The rank-and-file soldiers simply needed proper training, physical exercises and drill, and certain basic fighting techniques for battle. Training therefore revolved primarily around some basics such as: "how to salute, walk, and respect higher ranking people" (BC-KP4). These forces, however, needed to be made use of properly, using proper strategies to assign their tasks concisely: "Without proper training and clearly knowing all the military fighting techniques, and if they are not assigned to their tasks concisely, the situation [during battle] would be chaotic" (BC-KP4). Battle success rests upon two pillars: proper training and correct commands. While training and drills are what makes a recruit a good soldier, it is good strategic planning that makes a commander a good leader. The planning and strategizing is most important, so that soldiers do not just randomly shoot their bullets: "Shooting had to be proper, not simply random shooting. Therefore, the commander had to be good at strategizing. They had to know how to correctly shoot in the plateau and on the mountain. They had to know the rules according to the geography" (BC-KP4). Although there are no real winners or losers in battle, the plan is most decisive when it comes to losing men: "In battle, there are no real winners or losers. It is the plan that always wins. The difference is only that a plan eventually loses less or more soldiers" (BC-KP1).

Field Conflicts: Barbarians and Military Greenhorns
When talking about the impression new recruits and soldiers made when they joined the forces, the old military elite commanders recounted them as being rather chaotic, having long hair, and wearing make-up (BC-KP4). However, not only the lower ranks appeared to be barbarian; the upper ranks lacked basic discipline, and

the strongmen – who had no military leadership experience before going to the resistance zones along the Thai border – were greedy and illiterate according to the commanders: "He was an illiterate. He knew nothing. He was greedy" (BC-KP1). In the end, they saw in them greedy and brutish businessmen. Military regulations, rank, and the leadership of the commanders from the old elite seemed to count to nothing for these strongmen.

While these commanders residing along the border were deemed barbarian and greedy, others who joined from abroad were considered literate, even intellectual, but lacking any military knowledge. Their attempt to take over military organization and planning was greeted with much resentment and skepticism by the military elite circle. They even made them responsible for the split within the non-Communist forces and KPNLF, in particular, which led to the eventual failure of the forces. Normally, as one commander stressed, the KPNLAF was very good because "we had old civil servants and soldiers such as Grandpa Sak Sutsakhan, Grandpa Dien Del, Chhim Bunyong, Hing Bunthon" (BC-KP1). But then the split came due to a power struggle between the old commanders and military greenhorns coming in to take over military leadership and regulation. In their view, the main reason for the leadership quarrel was that the greenhorns let "new resistance forces in" who had no knowledge of "military regulations" but still wanted to be in charge of military matters (BC-KP4).

Power Practices and Institutions

Commanders from the old military elite largely followed military drill techniques they have learned at military academies with a French curriculum. While the broad war strategy is of primary interest, lower ranks are physically drilled and collectively humiliated. The lower ranks are essentially different to the rank of war strategists, and in need of physical conditioning. They need to adhere to military ranks and regulations by constant drill and stick to formal military procedures and strict timings. Drill served as the main force to form the formless clay that was the brutish nature of the belligerents along the border. As a result, the Cadet School served as the main institutional tool for shaping the movement according to the ideals of the old military elite.

Training and Physical Drill
Old military elite commanders are the ones who set up the military training center such as the Cadet School in the KPNLF. Before their arrival, there was no training school or any kind of training at all. Schools came into existence from 1980–1981 onwards, and Funcinpec also sent some of their commanders to the KPNLF Cadet School before establishing a school of their own (BC-KP4; RC-F1). Similar to the intellectuals, the military elite drew a clear line between the style of training used for the rank-and-file soldiers and that for commanders, each being trained at different institutions. The curriculum was also divided into two courses reflecting the gap between the lower and the upper ranks. At the lower level, in the rank-and-file school for recruits up to the rank of a section leader, a course

was taught that was designed for everybody who went through military training, and consisted of basic physical exercises called 'Tom 1': "The training consisted of crawling, crouching, shooting, jumping over fences, and running from place to another" (GC-KP2). One major pillar of this course was also to learn the basic steps involved in saluting and standing in an open space during a public meeting or before pledging allegiance to a flag. This was called *Somrak Prong*, meaning to relax and be ready (BG-F1). After 'Tom 1', there was also 'Tom 2', for mid-range commanders and beyond, which taught the strategies and tactics of guerrilla warfare, planning, and geographical and military preparations. Furthermore, the ranks at the top had the opportunity to further specialize in one of the listed areas.

Looking back on the effects of their own training and curriculum, the commanders believed that they considerably lowered the indiscipline and chaotic state within the forces that they encountered when joining from abroad: "As you could see [at the beginning], they had long hair and had make-up in their face. Since we encountered this problem, we had to establish a military school with General Dien Del" (BC-KP4). The intensity and duration of the training depended on the good will of the top commander and on the rank of the trainee. In theory, everyone was supposed to receive the same degree of training starting with 'Tom 1', then 'Tom 2' for mid-range soldiers, and ending in specialization at the top. But in reality, training was merely for mid-range commanders who received 'Tom 1' and 'Tom 2' and for some top-ranking commanders who joined with a non-militaristic background (since old military elite commanders drafted the entire curriculum). However, strategic and tactical training at the higher ranks was of primary interest for the old military elite anyway. Of course, lower ranks were to be drilled to behave orderly, but their main interest remained in military strategy and tactics. Soldiers only had to know basics regarding how to use a gun and find shelter in battle, but commanders had to know more and be able to direct it all (BC-KP4).

Rewards and Punishments
Similar to the intellectuals, the military elite's system of rewards and punishments focused on the soldiers' physical means, and promotions were given to recruits with a similar background, i.e., those with a military education or family traits similar to those of the commanding officer. In terms of rewards, the commanders from the military elite only gave a few special rewards to individuals but constantly focused on the 'basic needs' of soldiers that need to be covered. Similar to the intellectuals, basic safety was a priority when living with their commanders. Second, the soldiers needed enough food to eat. Third, they needed a shelter to live in. Fourth, they needed to receive basic training and maybe even some religious advice "to let them respect Buddhism and believe in Karma". And finally, they needed a job to do in order to be busy in any regard (e.g., BC-KP2). Thereby a new variety of patrimonialism was added to the multi-centric and relational practice. However, it changed its content and meaning within the military elite as well. If they were given, concrete rewards were in a militarist style and allocated during public gatherings, when individual soldiers were singled out to step forward and be complimented on their performance. In general, rewards and punishments were

restricted to these public gatherings when meetings took place with hundreds of subordinates. These morning sessions were central for instructing the lower ranks: "There were meetings every morning. Each morning at seven, we usually pledged allegiance to the flag. Then, the commander would instruct and morally educate the soldiers for at least 15 minutes" (RC-F2).

Similarly, punishments were centered on physical instruction during these meetings as well. Military commanders seldom sent their soldiers to prison or took revenge by killing a subordinate for disobedience. Disobedience was a sign of lacking physical drill; therefore they were slapped, beaten, and kicked in front of the others. Most effective, as one commander contended, was "kicking them or giving them physical punishments, push-ups, or any other kind we could think of" (BC-KP1). It was a very special leadership style as one subordinate remarked:

> These commanders had their very own way of collectively blaming soldiers. After the morning exercise, these commanders would blame some soldiers in the early morning at seven. The misbehaved soldiers would be called out to come to the front and blamed or penalized using soldierly procedures such as running, physical penalties, or the like. (SC-F1)

And everything had to be done in public, as one military elite commander highlighted: "They were complimented in public and they were also punished in public. We slapped them in public" (BC-KP1). For them, prisons did not work out as well as these collective humiliations and drill. Physical instructions and collective modulations rectify those who misbehave or conduct themselves in a 'disorderly' manner without acknowledging military ranks and regulations. Like for strongmen and intellectuals, recruitment for the military elite was based upon their own social background. Positions were given to these militarily skilled arrivals at the border who had been in a position of mid-range military personnel before and had experience in implementing orders from higher commanders.

Preparation and Combat Control
Before sending troops to battle, military commanders had to take care of a lot of details. Immediately before getting started, they had to select which troops to send. There were always a few who were not brave enough or were too willing to fight. These soldiers need to be sorted out right away:

> I would look at soldiers' faces before going for a combat operation. If I found out that someone's face was not good, I wouldn't let him go. Those with a too happy face wouldn't go either. Their faces had to be normal. If they looked too happy that would be a problem. If they looked too sad, they would be shot by their enemy in the battle. (BC-KP4)

But in the end, this was a very late stage of combat preparation and not central for the success of an operation: "The most important thing was giving the correct commands" (RC-KP2). Commands, larger strategies, and concrete tactics for an operation are much more important than whether the soldiers leaving the camp for battle have a happy or sad face. Therefore, the old military elite ran the third office within the general's staff, an office responsible for planning and strategizing combat operations.

No matter how motivated these soldiers were before battle, they would lose their motivation if the commands they received were unorganized or not properly timed. If the commander did not give the correct commands at the right time, with proper strategic consideration being made prior to engagement, they would lose too many soldiers and the rest would become "mentally weak" (BC-KP1). Therefore, the "most important trainings were the strategic and tactical ones for commanders" (RC-KP2). For that reason, these commanders spent months, if not years, preparing for strategic planning, tactical combat preparation, and field investigations. Success depended on proper investigations, preparation, and a good timing:

> Only if we clearly investigated [we could eventually be successful]. If we didn't know the battlefield most clearly, we wouldn't succeed. For example, when we planned to attack a barrack, we had to spend months or even up to years to do the necessary investigations, not just a few days. We had to know where the cannons, 80s, 120s, the DK, and the heavy weapons where, but the most important thing for each operation was at which time we could go to the interior. If we could do it and the time is right, we would be able to attack successfully. (BC-KP4)

The commanders had to know each and every detail and prepare exact packages for the forces: "They [commanders] had to know how many soldiers were going into one unit to fight, who would lead them, so that they could prepare the budget and propose that to the higher officials. Food preparation, who to be send, T.O. batteries, guns, medicine, and the like had to be ready" (RC-F2). To control the situation in the field and monitor whether a unit actually fought was to be done by co-operating villagers and patrolling units, who wrote reports on activities of others and whether or not they heard any gun shots nearby (BC-KP1). Furthermore, the commanders themselves had to write reports, as already mentioned within the intellectual commanders' section. These reports, however, were mostly busy work that did not progress through the channels to the upper echelon or were ignored by strongmen in action. And individual combat performance was a detail that completely slipped the attention of the military elite commanders. When talking about it, they simply pointed to the chain of command with subsequent reports on performance going upwards, acknowledging again that the main problem was that the upper echelons showed no particular interest in filing reports or being monitored (BC-KP1). The simple reason is that the military elite commanders, by contrast to the strongmen, did not join the battle and left it all to their subordinates to take care of the proper implementation of what they had strategized long in advance.

Intellectual Anti-Intellectuals

The status of the Khmer Rouge leadership is much more inherited than one might guess when thinking of a radical peasantry-based Communist revolution, and it even contradicts their own ideology of anti-feudalism, anti-intellectualism, and anti-capitalism. The very top, i.e., the members of the party's Standing Committee, had deep roots in the upper elite of the country for decades, overlapping political circles with the Democratic Party, and received a very high level of education (mostly at the Lycée Sisowath and at universities in France). Khmer Rouge leaders received an exclusive and elitist education, as did the intellectuals and old military elite within KPNLF and Funcinpec. They were members of or well connected within the Cambodian political elite for decades already, being introduced to Socialist thought as students in Paris during the early 1950s, and becoming a rather paradoxical circle of 'Intellectual Anti-Intellectuals' – a double-edged position that forms the very basis of their discourse and practice confronting the old intellectual elite. During these years, when the intellectual elite of the Democratic Party promoted them, they developed a 'heretic habitus', as Bourdieu would call it, competing with the habitus of the intellectual elite (if one takes the intellectual elite's discourse as 'orthodox'). On the one hand, they were up-and-coming intellectuals themselves and were – as they believed – in the position of superior theoretical knowledge. However, on the other hand, they propagated the primacy of practice that should be experienced without being spoiled by theory or 'intellectualism'. This paradoxical position results in the construction of 'normed practice' in all fields of their rule except, of course, within their very own social sphere.

Habitus Formation and Lifecourse

Almost all members of the Khmer Rouge top leadership shared a similar social background and lifecourse. Their biographies are well established by lengthy and detailed secondary literature and, in the case of Saloth Sar –better known by his henceforth-used *nom du guerre* 'Pol Pot' – by autobiographies written by various experts (e.g., Short, 2004; Chandler, 2000), as well as tribunal proceedings. Throughout their lifecourse, beginning as early as their schooling, they formed a close circle of friends and sponsors culminating in shared position in the Central and Standing Committee between 1975 and 1981, or the Military Directorate after 1981. Far from being an incorporation of their own ideal of purity in terms of anti-capitalist and anti-feudalist class backgrounds, most were connected with the leading elitist political sphere of the country, some even for generations already, with families serving in the king's court (e.g., the Thiounn family). Because his father was a well-known landowner, Pol Pot was maybe the closest to the party's own class ideal among all its leading members: "His father, Pen Saloth, was a prosperous farmer with nine hectares of rice land, several draft cattle, and a comfortable tile-roofed house" (Chandler, 2000, p. 8). However, in addition to the moderately wealthy position of his parents, the family was also well connected to the royal palace, with family members working as servants to the palace and, most notably, his cousin Meak being awarded with the title of *Khun Preah Moneang Bopha Norleak Meak* after giving birth to King Monivong's son Prince Sisowath Kusarak in 1926 and his sister Roeung becoming one of King Monivong's concubines as well.

But not only Pol Pot had some strong connections and social resources at hand. Many of the Khmer Rouge inner circles went together to the elitist Lycée Sisowath in Phnom Penh, finishing their degrees there or at another exclusive high school in the capital. The main reason, however, that they established themselves as an up-and-coming contesting intellectual elite was that they came under the sponsorship of members of the Democratic Party during the late 1940s and early 1950s. Contact with these intellectuals, mostly through family connections to some members in the circle, led to the allocation of scholarships to these graduates to continue their studies in Paris. Hence the 'who's who' of the Communist leadership went to France together and encountered Communist thought at a time when French Marxist student movements were on the rise. Many Cambodians even became members of the French Communist Party, while an infamous Marxist-Leninist reading circle was formed with central figures such as Saloth Sar (Pol Pot), Ieng Sary, Khieu Samphan, Son Sen, Thiuonn Mumm, Thiounn Prasith, Hou Youn, and Hu Nim as its members (Becker, 1998, p. 212). Some discontinued their studies because their scholarships had been cut for more or less political reasons (e.g., Son Sen, Pol Pot); others finished their studies with a PhD (e.g., Khieu Samphan, Hou Youn). Those with a PhD returned to Cambodia and came even more under the influence of the Democratic Party, who then tried to promote them in politics and eventually in the government. Until the Democratic Party officially ceased to exist due to political pressure in the *Sangkum* and on an informal level even far beyond, both circles remained closely connected, and some biographies constantly overlapped.

Until today, many Democrats who were involved in these circles back then are puzzled by the subsequent lifecourse of their former clients and their later political trajectory after the dissolution of the Democratic Party in 1957. At the beginning, the left wing of the Democrats remained sympathetic with the up-and-coming 'Communist Party of Kampuchea' (CPK). However, for more than a decade after their return to Cambodia, the circle surrounding Pol Pot took control of an already existing but marginalized Cambodian leftist movement, formerly led by Tou Samouth (murdered in 1963) and whose formation could be traced back to the split of the Indochinese Communist Party into three national formations in Vietnam, Laos, and Cambodia in 1951. Nuon Chea, who studied in Thailand rather than France and later on became known as Brother Number Two within the Khmer Rouge movement, was one of the few allies of Tou Samouth, who managed to be part of the new leadership (Murashima, 2009). During this time, however, he and his associate Keo Meas were officially deposed from party leadership, while Son Sen and Vorn Vet took their positions in the party's Central Committee. At the time when the 'Paris circle' took over in the Communist movement, it was barely worth being called a movement with just one or at best a few hundred members (for the history of the Khmer Rouge, see Etcheson, 1984; Kiernan, 1996; 2002; Chandler, 2008). Later on, however, the party announced that the birth of the movement was 1960 (when Pol Pot and his associates became high-ranking party members in a small conclave meeting) and not 1951 as stated before (when the 'Paris circle' was still in Paris and the ICP split), thereby separating or cleansing itself from its predecessors and their connections to the nationalist Khmer Issarak and, most notably, to Vietnamese comrades and patrons within the Indochinese Communist Party (Chandler, 1983).

Starting in 1963 and taking refuge from intimidation by the government in the 'no man's land' of the northeastern Cambodian provinces of Ratanakiri and Mondulkiri, the movement remained rather marginalized and without serious military capacities. At first it was more of a refuge for a small group of Communist intellectuals fleeing the Sihanoukian police than a mass movement with a wide peasant base. But the top leadership of the Central Committee found a shelter there, as did many former Democrats. After being accused of instigating a peasant revolt at Samlaut in 1968, Hou Youn, Hu Nim, and Khieu Samphan also joined the Communist cells operating in the remote jungles. Only because of the US bombings, setting in as early as 1965 and peaking from mid-1969 to mid-1970, and due to the coup against the king, who immediately agreed to join the Khmer Rouge in a united front, did the Khmer Rouge ranks swell considerably. Sihanouk called 'his children' to join him in the 'marquis', granting the Khmer Rouge a strong military force that was able to take power after the Lon Nol regime suffered from corruption and the pullout of US forces from Indochina in 1973. The US left a military vacuum behind, of which the Communist movement could make use.

The main point to be made regarding how this group formed its habitus, however, is their position as an up-and-coming and competing intellectual elite. A diachronic view on their lifecourse shows how closely connected they were to political and intellectual leaders in the country, but at the same time, they engaged in a symbolic struggle to differentiate themselves from these leaders. Within this competition, they developed an anti-intellectualism inspired by Marxist thought that was adopted in French leftist circles. However, their own habitus still remained one of intellectuals. They are intellectuals like others, but due to a symbolic struggle within their lifecourse and the influence of Marxist thought, they developed a seemingly paradox habitus of 'anti-intellectual intellectuals'.

Access to and Mobility in the Field

While the Khmer Rouge planned to erase exploitation, feudalism, and capitalism when they took power on April 17, 1975, the upper echelon of the leadership and most members of the party's Central Committee consisted of intellectuals of the 'Paris group'. Such rather moderate members with close connections to the Democratic movement and its old patrons of the 'Paris circle' were either soon purged or marginalized, as one former Democrat and long-serving comrade in charge of Khmer Rouge bank accounts during the 1980s and 1990s stressed:

> Rith contended that those most victimized by the regime were the petty bourgeoisie, especially Democrats who joined the revolution, because they wanted to exercise their freedom of expression, to put forward their views, but could not do so. Rith himself was in this category, but lasted a long time because he was seen as one united front cadre who could organize specific things with regard to foreign trade, things that only someone with education could do. (Interview Van Rith, 2003)

While the leadership was screening lower ranks for 'pure' revolutionaries in terms of class, the leadership itself was highly impure. Those with vital functions (such as running 'foreign trade' and bank accounts) may have survived if they proved they had a proper spirit and obedience to the party and if they refrained from exercising too much 'freedom of expression'. A high level of education and a feudalist and capitalist class background were reinterpreted as signs of awakening rather than being seen as unsuitable for a pure Marxist-Leninist revolution and its leadership in particular. A proper 'revolutionary spirit' may have at least in theory outweighed an impure class background (see Heder, 2012). Here education meant increased knowledge and clear-sighted analysis of society instead of being first on the execution list. The confession written by Thiounn Prasith is a perfect example of the paradoxical position in which these leaders found themselves. While omitting reflexive pronouns such as 'I' and 'me', Prasith confesses his sinful lifestyle rooted in his strive to obtain private property that he was living on just a few months before: "[My] class standpoint was not clear. [I] lived like the exploiting class" (CGP, 2010, p. 5). The reason that he wrote the confession was that he owned a house and some capital in France and was not willing to give it to the party, stating that private property mattered a lot to him. However, it was not only such ownership that made him weak. His emotional belonging to a French(!) woman, his wife, was rationalized as being deluded by his French feudalist education but stays unresolved until the day he wrote the confession:

> [My] realization at that time [during the mid-50s] was just loving the country. It had not yet reached the level of loving the class. Another thing was that my national spirit was not high, to the point where there was confusion [in ideas] between loving the nation and loving internationalism. Because of these reasons, I married a French woman in 1954. That matter was also because of the feudalist exploitation and the colonialist education. The wrong decision that I made enabled me to have complications within my feelings until today. (CGP, 2010, p. 3)

While belonging to different classes forms the minds of the people, the leadership believed that each individual could overcome his class with a 'nationalist spirit' and be re-educated into a proper revolutionary. This process of rectification was more difficult depending on the degree of class exploitation you 'enjoyed' before, but it was at least theoretically possible (only members of the Lon Nol state apparatus were deemed unfit for being 'reconstructed' into pure revolutionaries). Hence the leadership ranks refined themselves successfully despite belonging to highly impure classes (or, as in Thiounn Prasith's case, many *almost* refined themselves). While the Khmer Rouge's emphasis on practice and peasantry has been widely discussed, its combination and pervasion with intellectualism and scientific normativity, which is a result of this seemingly paradoxical intellectual position in the field and its reversion, has not yet been analyzed systematically. Yet the Khmer Rouge leadership's discourse and practice was ruled by what might be called 'normed practice'. Although this study focuses on the years after the genocide, it would be artificial to separate discourse and practice before and after 1979. The practices and ideas that changed have been highlighted already ('no killings policy' in particular), but the bulk of it can be found during both timeframes. And the position of the leading 'Paris group' stays the same until the very end (during the late 1990s).

Classification

As in all branches of Khmer Rouge ideology, the good revolutionary soldier was a product that is normalized in proper practice. In order to guarantee his perfect formation according to revolutionary standardization, he would ideally be a blank page – meaning an ideologically unspoiled child that could be molded into the current revolutionary model like formless clay. Accordingly, the party center claimed to be at the center of knowledge. Similar to other intellectuals, they were in the position of superior knowledge and diffused this knowledge down the party hierarchy. Each party member was allocated a certain share of knowledge that was necessary for him to perform his revolutionary tasks in line with his social and organizational position. Each level had a leader who knew slightly more about the revolutionary truth than those below him in the hierarchy:

> The DK leaders had their own crystal clear principles regarding who was to know what and what someone was not to know. This is a very important part of the story. It was not the case that everything was generally known. There was somebody who was leader at each and every level. (Interview with Van Rith, 2003)

Proper insight into the heretic knowledge increased with an agent's rank, while there was not much to be known at the bottom of the hierarchy.

The anti-intellectual intellectuals within the top command were those were 'clear-sighted' with regards to what practice should look like, how revolutionary minds are formed, and how the everyday life of each subordinate should be planned to the very last detail. The party's economic four-year plan fixed all basic necessities for the people:

> Material Necessities for the People
> • On a co-operative, family, and individual basis
> • Clothing – scarves
> • Bed supplies – mosquito nets, blankets, mats, pillows
> • Materials for common and individual use: water pitchers, water bowls, glasses, teapots, cups, plates, spoons, shoes, towels, soap, toothbrushes, toothpaste, combs, medicine (especially inhalants), writing books, reading books, pens, pencils, knives, shovels, axes, spectacles, chalk, ink, hats, raincoats, thread, needles, scissors, lighters and flint, kerosene, lamps, etc. (D00591, p. 111)

Due to superiority in knowledge, each superior in the chain of command was assumed to be correct in what he said and did, although those below him may not have understood his rationale or why certain things were allocated while others were not. The hierarchical structure of knowledge also became reflected in the idea of 'democratic centralism' within the Khmer Rouge. While 'democratic' may spur rather different notions in different contexts, it basically means that the minority, that is, the individual, had to accept the majority's decisions, that is, the collective (ECCC, 2012). 'Centralism', in turn, means that orders come from the collective's center (Angkar).

Practice and Anti-Intellectualism
Ideally, and the anti-intellectual intellectuals seem to follow Maoist thought here, practice should form knowledge without intellectual and theoretical interventions. Practice forms the mind of the people, and proper revolutionary practice forms proper revolutionary people. Within the heretic discourse, it is not the theoretical knowledge that guides the people but their very own practical experience. Thus, in order to become really knowledgeable, you should not learn at schools and obtain a Bac I or II but irrigate fields and build dams. Therefore, party slogans stated: "The spade is your pen, the rice field your paper" (Hinton, 2004, p. 96). However, as already indicated, still the party center decided what counted as *proper revolutionary* practice and which practice was already wrong and deluded by imperialist and capitalist thinking.

Build and Rectify According to the Norm of Collectivism
Those following the wrong practice and guided by individualist thought were to be corrected or rectified. Each subordinate had to act according to the political line. Combatants as well were to keep in mind that they had to follow politics within their martial practice. While keeping strong in revolutionary practice, they may have forgotten their actual capacity and material aspects such as the usage of weaponry: "Therefore, revolution and mentality are the two factors which determine our victory. But we must not look down on weapons" (D00389). While the social being of the subjects is formed by their practice, this practice became preset and normed by the Communist leadership in order to fulfill the Socialist ideal of pure revolutionaries. A pure Socialist mentality is free from individualism of any sort. As a result, the party slogan went: "Destroy individualism; built the collective spirit!" (Boua et al., 1988, pp. 193–4). Absolutely everything was to belong to the collective, and individualism was to be prevented: "We won't allow individuation to rise again" (Pol Pot, 1988, p. 182). Individualism and ownership were considered concepts of the old exploitive feudalist and capitalist classes.

As a result, the party center constructed a purely collective social system, in which each trait of individualism was to be eliminated. People had to be redirected towards the Khmer nature of collectivism, which became deluded by external forces importing individualist thought and exploitation. While most of the ideas are widely known, its application to the military organization is what is important for our purpose. The revolutionary military forces had to be constructed and rectified to fit a proper revolutionary mentality of practical experience and collectivism like any other social group: "Our forces could only grow strong if we maintained and built them up well. In order to maintain and build up our forces effectively, we had to constantly promote revolutionary mentality" (D00389). In contrast to the belief in the inherited nature of the social reality, the Khmer Rouge formulated and practiced a producibility and constructability of the social structures and, henceforth, the individual. This constructability was possible by collective organization according to a preset norm.

Doctrine on Class and Political Purity
Not everyone was able to undergo such a reform in thought with an equal likelihood of success. High-ranking members of the old exploitive class were heavily deluded, and others, such as former soldiers and policemen, were even beyond remedy. The mentality, thinking, and henceforth probable resistance against the revolution corresponded with the social class: "In Khmer society [...] each class has a philosophy, principles, and activities specific to each problem, large or small, and different from other classes" (Party magazine, Carney, 1977, p. 30). However, the level of delusion is set not just in social terms but also in spatial: Those living in towns during the Communist takeovers were regarded as "new people" or "17th of April people" who were influenced by individualist, royalist, and capitalist thinking. This made it rather unlikely that they would become pure revolutionaries, but not impossible. They still could prove their proper political attitude in practice while fulfilling the collective's or Angkar's tasks for manual labor in rice fields or in irrigation projects. But in order to gain membership to core organizations, their background was unsuitable, which is why no recruit in the military had a capitalist or feudalist background (other than those in the high command, of course).

The economic background, as measured by an individual's material belongings and standard of living before the revolution, determined the individual's level of delusion. But whether someone was an enemy of the revolution also depended upon his "political spirit" (cf. Heder, 2012). As a result, the party believed that a new and pre-structured social environment would rectify or 'refashion' the populace, since most were at least potentially good revolutionaries. Hence, as Nuon Chea elaborated in court, knowing an individual's economic background was important in order to "make a clear-cut distinction between friends and enemies (ibid., p. 6), while the 'political spirit' had to be assessed in order not to "lump everyone together" (ibid., p. 8). This, however, made it much less clear-cut to differentiate between friends and enemies. Even worse, those belonging to the class of 'base people', who lived under Khmer Rouge control before taking power in April 1975 and who were peasants in the countryside, could have a poor political spirit as well. The problem, however, was that this made the differentiation between friends and enemies far less clear-cut, as the common picture of an enemy signified: An enemy was described as a Khmer with a Vietnamese mind. On a phenomenological level, it was impossible to say who was an enemy and who was not. The only way to uncover counterrevolutionary spies was to observe people closely and judge their traitorous acts. When such acts were hidden behind a compliant face, constant vigilance and surveillance was needed to observe and analyze people's actions, as a cadre noted in his study book:

> We see the enemy acting already, but we say 'No matter'. This comes from our instruction is not yet hot [enough]. As [for example] when the enemy drinks palm sugar water, and then defecates in the drinking tube, we say he is lazy, [and] not that he is an enemy. In fact, he is an enemy. (Ea, 2002, p. 23)

Blank Pages

> If you wish to know how things happen, ask adults; if you wish to see them in a clear light, ask children. (Party Slogan, Locard, 2004, p. 142)

There was just one group that could be trusted for core organizations such as the military: the children of peasants. They were the epitome of yet-to-come social constructability and were therefore singled out by Pol Pot for leading the revolution into its future:

> Youth is a period of life in which there are very rapid changes ... It is a time when consciousness is most receptive to revolution and when we are in full possession of our strengths ... This then, is a general directive of our party ... It is the youth of today who will take up the revolutionary tasks of tomorrow. (Carney, 1977, pp. 48–9)

The youth were chosen to be integrated into the party's core. Party slogans praised the untouched and innocent nature of young people that still could be molded: "Clay is molded while it is soft"; and: "Only a newborn is free from stain" (Locard, 2004, p. 143). Therefore, children were perceived as blank pages that could easily adapt the scripture of revolutionary thought. Yet Pol Pot also stressed that this receptive nature of children meant that they constantly needed to stick with the revolution and stay faithful at all times:

> Those, among our comrades, who are young, must make a great effort to re-educate themselves. They must never allow themselves to lose sight of this goal. You have to be, and remain, faithful to the revolution. People age quickly. Being young, you are at the most receptive age, and capable to assimilate what the revolution stands for, better than anyone else. (Locard, 2004, p. 144)

As a result, even a child had to stay vigilant as to proper revolutionary performance and system of thought– a vigilance that was demanded by each and every one in the movement.

Normed Practice within the Military

For the top command within the Central Committee of the party, proper revolutionary practice was far from a simple principle of 'learning by doing' in combat. Documents from the leadership on how to plan combat operations and how to conduct military training are therefore highly formulaic in how they

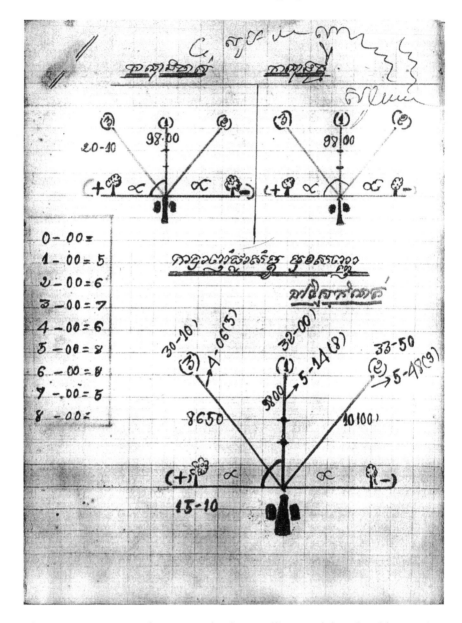

Fig. 4.1   Excerpts from a notebook on military training: bombing angles.
Source: Documentation Center of Cambodia.

meticulously prescribe a soldier's behavior and collective action. Far from being
anti-intellectual in this regard, the center was highly intellectual and scientific
in its prescriptions. At times, they were rather unhandy and may have hindered

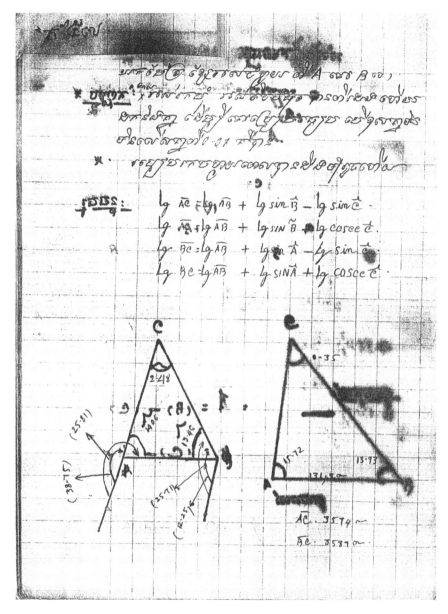

Fig. 4.2   Excerpts from a notebook on military training: logarithms. Source: Documentation Center of Cambodia.

unspoiled practice, as these pages on calculations of bombing angles and other logarithms from a notebook on military training sessions show:

Literally each step is normed like in an idealized drill situation. A Khmer Rouge "Guerrilla Training Manual" exemplifies various military techniques to be trained. While the list is too long to be cited, here are some examples that demonstrate how actions were regulated into their very last detail:

> 3) Queuing: Squad members come to the front and check whether they are in a straight queue or not. If yes, they shout to stop, and the soldier drop their hands at the same time, and then report to Group Chief or Group Deputy Chief the number of attendants-how many are present, how many are missing.
> 4) Right face: Before turning right with the body: Move with the left heel. The right heel with instep. And then move the left leg to join the right.
> 5) Left face: Before turning left with the body: Move with the left heel. Right heel with instep. And then move the right leg to join the left.
> 6) About face: Move the right leg backward with a 0.15m-space. Move backward and then join the right leg with the left.
> 7) At ease: Make a 0.5m-step with the left hand raised and placed on the belt in the front. Place the right hand down along the trouser seam. (Guerrilla Training Manual, 2003, p. 16)

Examples like this in study books are numerous and show how, similar to the marching formations detailed in Figure 4.3, each individual was kept in line via normed revolutionary practice:

The top command issued a declaration on the rules and duties of soldiers and commanders that were to be followed. Just few of these were mentioned in the field, except for these twelve essences of morality:

> 1. Must constantly love and respect people, labors, and farmers.
> 2. Must do your best to serve people no matter where you are.
> 3. Must not harm people. Even a chili or a bad word is not allowed to (steal or argue)
> 4. Must gently apologize to people if you had made any mistake on them and compensate them back what you had damaged.
> 5. Must talk, sleep, walk, and sit in a gentle manner.
> 6. Must not exploit women, regardless of any mean.
> 7. Must not gamble. It must be abolished.
> 8. Must not steal people's properties even a cent of money, a can of rice, or a single seed of rice.
> 9. Must always be gentle toward people, but hostile toward enemies.
> 10. Must always respect labors and farmers' sacrifices.
> 11. Must bravely fight against obstacles.
> 12. [Don't drink anything non-revolutionary in nature; Huy, 2003, p. 17] (D00382)

The top commanders constantly set up lists of different things, including what needed to be done, to be built, the twelve essences of morality, of revolutionary spirit, of winning the war, and so on. Then they set up norms in terms of fixed quotas for each one to be achieved, specifying when and how to do so.

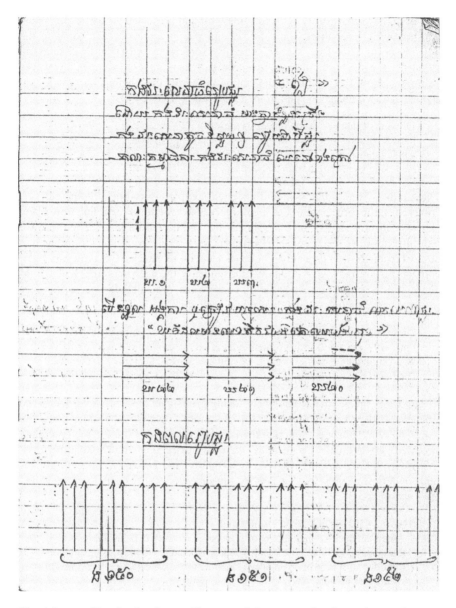

Fig. 4.3   Notebook for military training: march formations. Source: Documentation Center of Cambodia.

Knowing the Quota and the Correct Plan
The top commanders knew what needed to be ingrained by practice into the blank page of people's minds, and deviation became a sign of political unrest.

Part of this was that reality became something that presented no barrier for social constructivity. Everything could be set and put into practice according to the Socialist norm. Therefore, Khmer Rouge lists sometimes had a slightly phantasmal tone when it came to calculations for successful planning. For example, few weeks before invasion, the capital's radio station broadcast calculations on how to win against the Vietnamese in spite of their superior military power:

> In terms of numbers, each one of us must kill 30 Vietnamese. [...] We should have 2 000 000 troops for 60 000 000 Vietnamese. However, 2 000 000 troops would be more than enough because Vietnam has only 50 000 000 inhabitants. [...] [W]e would still have 6 000 000 left. (D30277, p. 43)

These quotas were unrealistic and operated completely on social constructivity. Oftentimes, the top command of the military also set quotas regarding how many battles each unit should have fought for the simple reason that: "The more we fight, the stronger we become" (D00389). As a result, continuous fighting was ordered from above: "Expand night attacks and quick attacks. Continuously fight, both during day and night" (D00389). Or suddenly, the guerrilla forces had to fight three times a day. A former battalion commander remembered one period very well due to the merciless quota placed on each unit:

> For me the period between eighty-five and eighty-six was unforgettable. At that time, we were asked to fight three times a day. It applied to every unit. If we could not fight three times per day, we were not allowed to get back to our place. (BAC-KR1)

At that time, the Khmer Rouge were losing ground on the battlefield due to the Vietnamese dry-season offensive. But the more you fight, the logic goes, the more you win. However, although the lists and directives from the top command appeared very detailed at times, they were also extremely vague and ambiguous. Everything had to be done vigorously, properly, and correctly, but nothing was said about how exactly to do it, leaving the subordinates with strict yet undefined tasks and plans:

> [How to] Control Cadres and Combatants
> - About the quantity and quality of leadership: In order to be in full control of the quantity and quality of cadres and combatants, current assignments and trainings must be properly managed in order to promote fighting forces and to ensure task completion.
> - Must be in full control of the number of people in the unit.
> - Must be in full control of the political situation, mentality, living condition, materials, capacity, techniques, strategies, commanding and fighting abilities, and skill of cadres and combatants in the unit.
> - Must assign cadres and combatants correctly and properly. (D21528)

In the end, everyone was to be vigilant in order to fulfill the will of the center, although it remains rather unclear just how this vigilance was supposed to be demonstrated in many cases.

Power Practice

Disciplinary and pastoral powers characterize Khmer Rouge power practices. Both power types normalize the subject by positing an optimal model that was constructed on terms of a certain result, the Socialist normativity. However, the operation of disciplinary normalization "consists of trying to get people, movements, and actions to conform to this norm, and the abnormal [is] that which is incapable of conforming to the norm" (Foucault 2009, 85). In disciplinary power, the subject is under constant surveillance, and corrective pressure is to be modified according to a preset norm. Thus punishment is not binary, as in sovereign power, but becomes essentially corrective on a continuum: "To punish is to exercise" (Foucault, 1991, p. 180). Thereby the subalterns can be made or transformed into a norm by constant rectifications – or they can fail and be classified as abnormal.

While disciplinary techniques transform subjects by meticulous and constant surveillance combined with 're-educative' practice, pastoral power uses the technique of confession to modify subalterns according to the will of the ruler. The confession aims at the renunciation of the self in favor of the leader's will (Foucault, 1981). Similar to disciplinary power, pastoral confession aims at a transforming and modeling the individual subaltern, in which wrongdoings and shortcomings have to be acknowledged in order to change behavior and thoughts in a next step. Thereby the confession constantly mortifies the 'deluded' self, as the self has to acknowledge its delusion and change to adapt to the preset norm. Both techniques have in common that they set a norm and establish control and obedience by constantly reworking the self. Likewise the Khmer Rouge discourse had an abundance of phrases that called for the deviant and deluded subject to be "rebuilt", "reworked", "redirected", and "corrected" to incorporate purely collective practice. Each agent is perceived as a yet-to-be ideal in need of rectification.

Setting the Norm: Rules on Collective Behavior and Class Purity
The norm of the Socialist top leadership is collectivism and class purity in each social exchange and each behavioral trait. While some rules are rather obvious, such as a uniform dress code and normed haircut for everyone, collectivism and class purity as a rule pertain to all aspects of disciplining soldiers. Recruitment, rules for promotion, and the system of rewards all reflect the Socialist ideal with which every soldier had to comply.

Recruitment and Promotion
In contrast to the KPNLAF and ANS, the NADK forces screened to determine if combatants fulfilled the norm. The Khmer Rouge top command was far away from an 'ocean policy'. Admission to core party organizations was granted only when an individual met three conditions:

1) Be active in struggling to perform one's tasks and having gained favor and recognition from most of the masses as a result of completing one's tasks.

2) Be of good caste […]
3) Be clean in terms of ethics and livelihood, and be politically clean and clear-cut. (Notebook of comrade Iv Bun Chhoeun, 2001, p. 12)

Furthermore, each potential cadre needed to be "introduced" by (in theory) at least two older cadres. Before entry, each applicant had to fill out a lengthy questionnaire on his life, his family, and his political stance to evaluate the three points mentioned above. Military directives state:

> Cadres on each level must be known clearly:
> Background
> Ability
> Intentions of each individual in the unit (D21528)

But in the end, the 'political spirit' and activeness in fulfilling all tasks given by Angkar were most decisive for promotion, as Gregory Procknow highlighted: "For the ones who demonstrated good work, the Khmer cadres would note this and recommend these selected few to mid-cadre level leaders that certain workers should be considered for promotion to the party leadership ranks" (Procknow, 2011, p. 141). Entrusted with the ranks within core organizations, however, were only those who were deemed pure in all terms: background, ethics, and political stance. Candidates for the Communist Youth League, for example, had to be most resolute and active:

> The Party's pioneer element is characteristic of the most courageous, resolute, vigorous, and active natures in combating, implementing political guidelines and making decisions on all kinds of work such as the revolutionary struggle, army, politics, economy, culture, social affairs, defense, correspondence, transportation, other skills and specialties, etc. (Notebook of comrade Iv Bun Chhoeun, 2001, p. 12)

The result was that the upper ranks, up to the rank of a brigade commander interviewed for this project, were filled with former sons of farmers who without exception had been recruited as children between the ages of thirteen and sixteen. For these child soldiers, ranks could be rather permeable, which is why they could gain comparatively high positions. But in the end, they could not rise into the inner circle at the top, which was largely occupied by the anti-intellectual intellectuals. However, lower- and mid-range up to even upper ranks melt within the Communist organization due to the strict screening using biographical questionnaires, the favoring of children, and a chance to constantly rise in rank, starting as a rank-and-file soldier and ending right before the Central Committee.

Cooperatives, Food, and the System of Rewards
The Khmer Rouge leadership constantly stressed the need to be a good example in collectivism and revolutionary conduct. Similar to other fields of activity, soldiers dined collectively within their units (or cooperatives). Everyone was prohibited

from owning anything, receiving money, engaging in any kind of trade or money-making, or marrying too early – i.e., at an age when they were to focus on serving the collective:

> [C]ombatants do not receive wages, any petty trade or money-making activity is prohibited, private property is not permitted beyond the essential necessities, and combatants, and all the more those who, because they are responsible for commanding them must set a good example, are advised to marry as late as possible. (Peschoux, 1992, p. 76)

As a result, goods were allocated daily "to everyone regardless of whether that person had any previous achievements" (BAC-KR1). Rewards were always collective (more food for everyone, for example). The only improvement a soldier could hope for was to be promoted to a higher position, which would mean he could be part of a collective that received better rations and were granted more rights. Although this also meant that they were regarded as politically purer, soldiers who were promoted to higher ranks also came under the closer observation of the top command. Due to constant purges within the party's ranks, this could be even more risky than staying at a comparatively low position. The constant threat of rectification, however, was present on each level.

Rectification of the Self – Part I: Disciplinary Techniques
The first disciplinary technique of disciplinary power was, of course, military training as a means of forming the 'formless clay' that was new recruits. Here, as the theory goes, practice was to form the clay, but practice was, in turn, to be structured along revolutionary theory and follow politics:

> The slogan is: Training must be done in accordance with the request of the fighting, and with the actual situation in the battle. Theories and practice are closely intertwined. Training and practice must be done simultaneously. (D21528)

While in practice, as Procknow highlighted in his study on the training implemented during the years of the Khmer reign, "[t]raining was largely undertaken in the form of hands on practical work experience" (Procknow, 2011, p. 140). The idea was to set up training exercises that simulated actual combat and then adjust them according to evolving factors based on subsequent battle experience. The distinction was not between practice and theory but between properly theorized and 'revolutionary' practice and incorrect practice. As a result, instead of 'learning by doing', some commanders within the higher ranks received much too realistic training that was preset to mirror actual combat events:

> The trainings during that time were first to march lock step, second to attack targets of Lon Nol sites, and third shooting exercises. There were instructors who were responsible for the trainings. The exercise was not a simple game. It was like actual fighting. For example, a village was used as the headquarters of

the military, the other two hundred soldiers had to patrol the cannons, machine guns, etc. It was like in reality. The soldiers had to do extensive surveillance since Khmer Rouge soldiers would later come for investigation. If they could start the battle, they would immediately attack. If not, they would have to retreat. Sometimes the training lasted from two weeks to one month [...] After six months, we proceeded to the real battles. (BAC-KR1)

The corresponding training manual states:

Training aims at achieving results within battle and gaining victory. Therefore, details within the trainings must be linked with the actual realities within battle. These realities include:
• Reality of the enemy situation
• Reality of the geographical conditions
• Reality of the weather
• Reality of our military actions (D21528)

But mostly the mixture between practice and theory was laid out as a test for the commitment of the new recruits by throwing them into battle:

When I first became a soldier, I was not given any weapons yet. They wanted to know first about our commitment, and second our attention. They wanted to know about the way we lived, the way we ate, and whether or not we could bear living in difficulties. I was asked to carry only one rice pot. (BA-KR1)

To carry on with assigned tasks during combat, despite having no weapon and only a rice pot to defend one's self, proved the recruit's commitment and work performance under difficult conditions.

Re-education and the System of Punishments
Most offences among the combatants' ranks were met with re-education at first (due to the pure class background), and if the 'political spirit' proved impossible, they were solved with execution. The party leadership believed thoughts could be reformed by manual labor. The deluded combatants likewise faced punishment by re-education:

As for errant cadre who had no fighting spirit and combatants who misbehaved, who broke the strict discipline both on and off the battlefield, there was a place for those removed from the ranks: Voat Kandal, where such liberals (serei) were sent to clear land and plant vegetables with a diet of rice and salt and re-education. They were then given the opportunity to volunteer to go back into combat, but if they made mistakes again, they would be returned and put to work at the Economy Place, in Prek Lovea, where the rice mill was, at the pig-raising unit on Phnum Veang, or at the fishing grounds. Most of these ill-disciplined elements couldn't resist booze and women. Rith claimed none were executed, declaring those involved in fishing got fat because they had all the fish they could eat. According to him, this was the only way to run a people's army, because if combatants were ill-treated for just minor behaviors, the parents of peasant boys would not allow them to go into battle. For this reason, the same kind of re-education was applied to those who deserted in combat. (Interview with Van Rith, 2003)

While the official code of law was strict and demanded pure collectivist sacrifice, the list of 'twelve essences of morality' mentioned above was constantly referred to by respondents. Moreover, several sources reported on a more concrete list of offenses that accounts for a deviation from the Socialist model (e.g., the Lawyers Committee, 1990). Here, medium offenses with a prison term of six weeks up to four months were reported for voicing critique, listening to the radio broadcasts by KPNLF and Funcinpec, flirting, trading with Thais, refusing orders (such as carrying ammunition to the frontline), arranging marriages for your own children, demanding pagodas and Buddhist ceremonies, believing in superstition, and demanding traditional wedding ceremonies. Heavy offenses were rather fuzzily distinct to moderate offenses and included voicing critique against the top command, voicing complaint about working conditions or anything else with 'political implications' with members of humanitarian organizations, moving between the zones, attempting to defect, murder, or endanger others, and, on very top of the list, stealing or selling ammunition and weapons. While there was one place for the moderate offenses (called Ta Sum's Place or Prison 80), serious offenses were dealt with at a separate prison (Ta Chan's Place or Prison 81).

While each unit maintained a close grip on surveillance of the individual combatants, there was also a "research unit" (*akeak pinet*), consisting of thirty to fifty people, whose commander was only responsible to the brigade commander. This unit's job was to ensure discipline, investigate wrongdoing, interrogate suspects, and ensure their re-education. Multiple inspection units checked upon each division internal affairs, discipline, and activities. These units moved from unit to unit and reported to the leadership, often operating in ordinary uniforms with no particular sign of distinction in order to mix easily with the troops. Others operated in ordinary green fatigue with a red armband bearing in white the three letters do-yo-ro followed by the number 87 (cf. Peschoux, 1992, p. 65). Moreover, each combatant had to report to the group leader, who reported to the section leader, and subsequently had to hand daily reports up the chain of command. Combatants had to incorporate the watchful eye and adjust their own thoughts and actions to be safe.

Rectification of the Self – Part II: Pastoral Techniques
Similar to the years in power, sessions for critique and self-criticism, also called 'livelihood meetings', were held in each military unit. Here combatants and commanders were expected to criticize mistakes and also implicate their comrades. Most of these sessions focused on the 'laziness' displayed by combatants in fulfilling their tasks. A report on a combatants' session read like this:

The weak point while fighting in battle was that soldiers were slow. Some soldiers were slow, walked away from their post, and could not meet the higher commanders' command on time. Soldiers retreated too slowly and did not prepare the materials well. There was still carelessness. Some people did not listen to the commands while pulling out guns. Secrets were not kept confidential. Those responsible for cooking also lacked preparedness (for immediate retreat/ departure). (D21528)

Confessions were regarded as fostering reconstruction through education, although they were merely confessions that included repenting and expressing regret, asking for a pardon, and declaring the intention not to do it again. Since these meetings were held each day, there was a constant circulation of shortcomings and rectification, following the logic that there is no logical end to the process of normalization, only asymptotic approximation. For the party center, reconstruction through critique was necessary to defend the collective and the Khmer race:

> After the discussion on work, there must be constructive criticism, self-criticism session as before. As for the leaders, we have a somewhat deep constructive criticism. For those below the leaders, we have a light criticism according to their awareness. As for the great people, we also have constructive criticism. We want to have solidarity in the organization. In one word, there must be an understanding in depth, this means that the politics is the front's responsibility, as for the spiritual concept and the executive framework, there must be solid cores in order to lead the country and the people to fight against the despicable Vietnamese enemy at the present time, to defend the country, to defend the people, to defend the race from now on. (Khmer Rouge file, 1982, p. 150)

These daily meetings were intended to 'alert' soldiers as to whether they were complying with the Socialist norm and to motivate them to strive to do their best within their duty. A former battalion commander put this 'alert' in rather positive terms:

> When we had strong commitment and clear attention, we could go anywhere and do anything. Therefore, a meeting was held every day in the unit during the Democratic Kampuchea era. The meeting was conducted to address the internal problems in the unit such as the inactiveness of the soldiers. If any problem existed, it would be mentioned during the meeting. They would raise many reasons to tell us. It was like an alert aiming at waking us up so that we would try harder on our own later. This was really an important thing because every day was meeting day. Normally both, you and me, when we do something wrong and got blamed every day we may get annoyed. When we get sick of being rebuked, we will work hard. This was done on a daily basis. (BA-KR1)

The practice of criticisms made combatants distrust one another and isolated them from their group. That might be the main reason why people 'happily accepted' critique. Everything else would bring them into trouble, since those who oppose are enemies of the collective and its new order. Sorya Sim and Meng-Try Ea give also an example of a meeting as reported by a former cadre:

During the livelihood meeting, Sok always talked about how late I was in getting up and how lazy I was. I was very afraid of everyone, especially Sok. I did not trust anyone. Everyone tried their best to search for one another's faults. I was working and living in fear and horror. I kept trying to work harder and harder, and I kept my mouth shut all the time. [...] The livelihood meeting was also one of the dangerous things because we were found and killed sometimes for only very minor mistakes. One of my fellows was captured and killed after his fault was raised by a comrade during the livelihood meeting. [...] When the group chief finished speaking, the young comrades had to stand up one after another. For example, first I started talking about my weekly activities; next, now I am finished talking about my weekly activities, and what I have done wrong, and now I am very glad to listen to all comrades' comments on my weak points. (Ea & Sim, 2001, pp. 23–4)

Confessing shortcomings in public strengthens the leadership's control of the individual group members. The subject had to acknowledge his shortcomings and rectify towards the norm, rework himself under the threat of being killed if he or she failed. The problem with the meetings was that people had to balance their critique: Criticizing yourself without being implicated in major counterrevolutionary acts and thought, and criticizing others without retaliation for it later on. Even more difficult was the fact that anything, no matter how big or small, could gain major significance:

A lack of speed in executing a task demonstrated a sabotage mentality. Zeal was an undeniable sign of longing for power. Thought has a much value as action. To be on the right track, one had to imbue oneself with the ideas of the party in such a way that the mind was perpetually mobilized to the party's service, without hesitation and without wasting time, like a machine. (Picq, 1989, pp. 107–8)

Constant critique was not just 'annoying' but also carried the risk of execution if the subject did not adjust himself and his actions according to the party line and prove willing to fulfill its duties and tasks without any sign of complaint. From the moment they filled out the biographical questionnaire during recruitment and through each and every 'livelihood meeting', cadres constantly had to emulate those thoughts and practices which best displayed the proper revolutionary consciousness. Their demonstrations of their innermost and revolutionary thoughts and attitudes determined their position in the party, their rights, and, most importantly, their personal safety. The same process of constant remodeling and rectification applies to the preparation for and review of combat events. Ahead of each battle, group meetings were held on how to implement the operation. After the operation ended, another meeting was held to review the shortcomings of individuals and the implementation in order to improve everything before going into future battles (cf. Huy, 2001, p. 20).

Panoptic Vigilance
The system of omnipresent and close-knit surveillance yielded a 'panoptic effect' (cf. Foucault, 1991), in which people constantly felt they were being watched.

A famous saying noted that Angkar had the eyes of a pineapple. You could not evade its glance; there was nowhere to hide, and no one to trust. Combined with the transformative pastoral techniques, everyone had to be constantly vigilant and observe his or her environment for even the slightest signs of possible danger. Everything was under Angkar's glance, and everything emanated from its powers:

> All revolutionary laws and regulations were promulgated in the name of Angkar; all transgressions were known to and were punished by Angkar. Angkar was everywhere, a pervasive presence that none could escape. 'Angkar has more eyes than a pineapple', the cadre said. Husbands and wives spoke of Angkar only in private, in a whisper, fearful of being overheard. No-one criticized Angkar in public; even the most minimally critical passing allusion could be enough to ensure arrest, interrogation, and subsequent disappearance for re-education. Danger was ever present; at no time did one know whether the spies of Angkar were listening. (Stuart-Fox & Ung, 1998, p. 54)

Angkar watched its subordinates from every angle, via each cadre, each comrade in the unit, and possibly even your own family members. Control became 'total' in every sense of the word. You could never know who reports to whom and what. Therefore everyone was in a state of 'hypervigilance' as described by Peg LeVine when citing how people tried to figure out rules and regularities for possible dangers that may arise. The slightest gesture or rhythm may serve that purpose, providing leads on people's intentions and plans:

> I studied the sound the soldier's footsteps and was able to hear the step that came to kill; they never really told us clearly their plans. One of my friends was called out in the night and the next day I saw his (severed) foot on the ground; the next night another friend was called out and he went to his wedding. I looked for small signs so I could guess better what would happen if I got called out. When we were leaving the city after April 17th, everything was confused; but when then over time we knew that if we saw or heard this and that, then that would happen. (LeVine, 2010, p. 10)

Each cadre in the security apparatus had to fear harsh punishments for even the slightest deviation from the norm as well. Members of the security apparatus, which was under control by the military, had to fear being categorized as an enemy and to face the fate of those they had been in charge of before. Cadres from the central prison S-21, for instance, were arrested on a regular basis:

> S-21 personnel were also arrested and either sent to S-24 [an attached center to Tuol Sleng] for re-education, or imprisoned in S-21. Cadres could be sent to S-24 for minor offences, especially when someone they knew was detained at S-21. For more serious offences, such as allowing a prisoner's escape, death or suicide before interrogation, the person responsible was considered as a traitor to the revolution, and was arrested. However, some witnesses suggest that the majority of S-21 staff members who were arrested, specifically those from Division 703, had not actually committed a serious offence. (ECCC, 2000, p. 114)

Everyone in the security apparatus and the military knew what it meant to be categorized as an enemy who had to confess under torture that he was spying for the 'CIA', 'KGB', or 'the Vietnamese'. The panoptic principle of rule leads to an incorporation of control or an internalization of the watchful eye, in which everyone acts *as if* being constantly watched. Angkar's eyes may not have been present at all times, but people had to fear their constant possible presence, thereby acting in constant conformity and taking over the duty of their own subjugation under the disciplinary norm and Socialist modeling. This corrective modeling knows no logical end; failures and deviations running against the purity of the model may resurface at any time.

Chapter 5
Military Operators

Many of the mid-range operators came from the patrimonial network of the leadership and reconstituted an entourage that was in place before the Khmer Rouge took over in 1975, thereby reconstituting long-standing social differentiations. Securing a position within the ranks of a company up to the rank of battalion commander depended on one's ability to reactivate social resources and, to a lesser degree, on the scarce cultural resource of having at least some military education. Furthermore, everyone interested in a job within the official camp economy had to attend and graduate from the Political Warfare School in the KPNLF camps. Membership in the respective political organization running the camp was essential for getting a job, getting promoted, and oftentimes for making a living. After attending the Political Warfare School, military recruits attended the KPNLF Cadet School. While at the beginning ANS soldiers were sent there as well, the royalists set up another school for themselves later on or sent their commanders to training programs in Thailand and Malaysia. Although they largely were clients of the leadership, not everyone was keen on joining the military resistance. Some had to be pushed into obedience. As one company commander highlighted, he refused to join at first "until one day when [name] told me that if I do not go, I would have to dig the land to burry myself" (CC-KP1). At times, social resources proved to be a double-edged sword. The first habitus group consists of this entourage showing high degrees of communicative and practiced loyalty. They had a medium level of military education either before joining the resistance, such as the old military elite (implementing their strategies), or they joined the Cadet School.

Loyal Tacticians

A moderate level of military education was the main resource for this group, which included tactical implementers of strategic plans as well as a patrimonial network, of which they were part. Commanders from the mid-range still felt the need to distance themselves from the ordinary soldiers in order to not to fall into the category of the rank-and-file soldiers. For them, military hierarchy and regulations were most important, and sticking with unquestioned discipline was demanded of themselves as followers of the upper echelon as much as it was required of their subordinates. None of these mid-range commanders made a career within the resistance, but all lingered at the same level from the beginning until the end of the war. Even today, most of them are still loyal to their former commanders and hold mid-range positions below them in peacetime. Either they were clients of old military commanders who already joined the military under Lon Nol or they went through various trainings while serving an intellectual or military elite commander within the resistance. Some also followed defecting commanders of the Heng Samrin government after having been to Vietnam for training. As mid-range operators, most would describe themselves as 'advisors', 'consultants', or simply those who implement directives given by the top command as military operators rather than as 'commanders'.

Habitus Formation and Lifecourse

Some of the mid-range commanders came from military families with a father serving as a mid-range commander under Lon Nol or the king as well (e.g., BA-KP2). They carried on an inherited tradition and family position. However, most of the 'loyal tacticians' came from rather well-positioned peasant families who could afford to send them to school instead of needing them to help in the rice fields back home. Although they were poor, they could attend school. Some of them, however, remember this life as the tough life as 'ordinary people':

> During these years, ordinary people faced a really tough life and school was far away from my house as well. I had no bicycle to go to school because we were too poor for it. There were only few bicycles in the entire village. Those who owned a bicycle back then may be compared to those having a car today. (BA-F3)

Many went to a pagoda in order to be able to attend a school and have a place to stay close to a school if bicycles were not an option and the home village too far away (e.g., BA-F4). Others were slightly better off as peasants, and they remember life as a student more pleasantly. Youth is remembered as being a life "under my parents' supervision and protection. What I had to do was just eating and studying" (BA-KP2). In the end, neither of these commanders received a high level of education. They either dropped out shortly before graduating with their Bac I due to the Communist takeover or they joined the military due to problems with or right after finishing Bac I: "I had to stop studying after grade seven as my living condition was really severe, my dad's leg was poisoned, my mum was pregnant, and I was the oldest among nine siblings" (CC-KP1). Most, however, were very keen on highlighting that they were literate either due to their studies as a monk or because they learned to read and write during their training as soldiers:

> I am a person who likes reading books. So far, I have read around three to four hundred books. I read everything that's in front of me. I reduced my reading only after I became old, but I still read at least a few pages per day. (AB-F)

While the military was a good option for these pupils who saw no alternative in the narrow job market, it was also the atmosphere of war that drove them to join the resistance. Some simply joined for protection during wartime. Having a gun and a uniform gave some the confidence they needed to move freely, particularly for the young pupils:

> Becoming a soldier meant that we could carry a gun. Even though we may not be allowed to harm anyone, no one would dare to harm us as well because they were afraid of us. So I decided that I wanted to own a gun to protect myself. (BA-F3)

The same commander even compared his uniform to a protective tiger skin: "When becoming a soldier, we were able to have a tiger skin to put on. We were wearing military uniforms and could move around freely" (BA-F3). At times, it seems, the military was the least dangerous place.

Access to the Resistance

After they went into hiding during the Khmer Rouge regime, none of the mid-range commanders interviewed became a member of the resistance during its founding years, which is why they were not part of the inner core that was elevated to leadership positions early on (like the strongmen, who rose heavily in position during these years). Right after the Vietnamese invasion, most first went back to their villages or to the city from which they originally came. Here, news reached them that their former superiors needed them at the border, or they even joined the incumbent state military together with their long-time bosses. Others went to the border but applied for resettlement in a 'third country', only until former superiors demanded their service for the insurgency. For many, it did not really matter which faction they fought for as long as they served their patron. Sometimes, however, their recruitment was not all voluntary and needed certain pressure.

In the end, all of the tacticians were loyal to one patron, and all received years of training, either already as a Lon Nol soldier, within the incumbent government, on training missions in Vietnam, or after joining the resistance. Loyalty and military tactical knowledge were central to their discourse and power practices, although some of them may never have received a formal rank. For them, the reason for rising to the rank of a mid-range operator was twofold: "First, I was disciplined, and second, I was smart with regard to military matters" (BA-KP2). Thus, the main resources of the tacticians included their tactical knowledge as trained and sometimes long-serving mid-range commanders and their unconditional loyalty to their commander, which was framed as military discipline and adherence to command.

Classificatory Discourse: Tactical Knowledge and Military Discipline

A good commander knows how to implement war strategies, while a good soldier adheres to military regulations and is trained like a knife: "We had to sharpen it. They needed to know how to shoot and how to fight properly" (CC-KP1). While the body of the soldier needs to be sharpened, it is the mind of the tactician that needs to be formed by knowledge in order to be able to decide on a proper battle plan:

> There are many factors [making someone a good commander]. There is
> (2) knowledge and decision. What do I mean with knowledge? We had to know
> the area. Second, we had to know our enemy's forces. Third, we had to know the
> types of weaponry and instruments our enemy might be using. Fourth, the most
> important thing was that we had to know where our enemy's leaders studied,
> and how many times the leader had been involved in battles already. Fifth, we
> had to know about our enemy's communications. We had to know where their
> supporting forces were and whether our forces were comparable to theirs. And
> then there was what is called the commander's decision to formulate fighting
> plans. Once we obtained clear information, we had to look at our side. We
> looked at how many soldiers we had, how much weaponry, what our supporting
> units were, how we are going to move from one place to another, and how much
> time would be needed and how soldiers should run. (BA-KP2)

Hence, a central component of being a good mid-range commander was proper
investigation for drafting tactical plans. For this, they had to "obtain clear, accurate
and well-refined information" from the field and regarding the state of their own
military units (BA-KP1). Most of these mid-range tacticians constantly used
French terms to describe the importance of making a plan, due to their training by
old French-educated military elite commanders.

In the end, battles were like a boxing match. One interviewee explained: "We
had to know for how long that guy [a certain commander from enemy with his
men] had been in boxing already so that we can proceed further. Basically, I think
it wasn't much different from boxing" (BA-KP2). However, he also stressed that
there was one major difference to boxing: You had to think first before moving.
Military education and training had to be well thought through before engaging in
immediate action:

> Those military trainings helped us to broaden our ideas. Before doing anything,
> we had to contemplate well. We were advised to think first before taking any
> actions. We could not simply launch an attack as we wished without thinking
> first. We had to think first. (BA-KP2)

In the end, contemplation was the key to winning a battle: "Leaders always had to
think after each combat. They had to analyze why they won and why the lost. Then
they jotted them down" (AB-F). But in the end, one had to engage in practice in
order to see what happens to a plan and to make changes, if necessary:

> I believe in a theory that those who have no mistakes are those who never
> engage in practice. As long as you are actually doing something, you will
> do some mistakes. This means we should better choose to do something and
> make some mistakes than doing nothing. Reasons could have been because we
> lacked bullets or forces. Or we didn't take enough care about our patients and
> so on. (AB-F)

While scripting plans was essential, a good commander had to rework his plans
constantly due to emerging factors that arose in concrete battles:

There are many unexpected things happening during combat. Ordinary people who are not exposed to these experiences do not have sufficiently strong skills in critical thinking. First we thought that as long as we could have enough food, we could sleep happily without our wife and children. But it was not like what we thought. New problems always emerged. We could try to solve as much as possible and would write down what we couldn't solve. We had to fix these situations, be in the morning, afternoon, or at night. As a commander, we constantly had to think during combat because there are always soldiers being killed or getting injured. Normally, people were so scared of the word 'death' that they did not want to go to combat. So what could we do with those who were still alive? This was a question. (AB-F)

The tacticians were occupied with the concrete implementation of and any problems related to the broader strategy. Central for getting a position, however, was to respect and adhere to the command. Referring to a Khmer saying, a battalion commander stressed obedience to military command as a necessary factor for getting a position. But, at the same time, it must not be confused with blatant appeasement:

Our Khmer saying says that a position derives from respect, commitment, alignment, and adherence to command. Appeasement might lead to danger someday. Normally, those who liked to appease others were not the ones who were serious with their work. They only wanted to please others. (BA-F4)

A good soldier who fulfilled his allocated position in the command structure within the military collective did not do so for his own personal profit. Military rules said you should obey your commander even though you may never be promoted or rewarded other than with your own death as a heroic sacrifice (BA-F2). This fits with the fact that most of these clients never left their posts or sought to improve their position below their commander. Their adherence to their own position while striving for nothing more than its preservation becomes interpreted as a soldierly honor.

A soldier's duty is very different from an ordinary man's duty and even more compared to Buddhist teachings by monks. A soldier has to fight and follow his duties in combat. He cannot ask first with whom he is fighting and decide later on whether or not to shoot. He needs to be quick and determined – otherwise he himself could be shot:

What the monks said was of course useful, but they themselves did not go to fight in a battle. While firing, the bullets do not recognize who is who. If we do not shoot others, we would be shot instead. So we have to kill others in order to avoid being killed ourselves. We do obey to monks' teachings, but the teaching itself was simply not applicable during battle. (2) If we only cared about those teachings, we would be shot dead. (BA-KP2)

In practice, military discipline does not allow for many ethical considerations. Buddhism makes sense and helps in certain areas, but it is not practicable when fulfilling a soldier's duty. The problem for the tacticians, however, was that ordinary soldiers were extremely loyal but undisciplined in terms of good conduct.

> Most of Khmer's rank-and-file soldiers were illiterate because they all just survived the Pol Pot regime, which is why they did not receive much education. Some didn't even know the letter [*kor*, first letter of the Khmer alphabet]. They even knew how to sign or thumbprint only after being instructed how to do it exactly. Because of their illiteracy, they believed in everything. Such kind of people only followed their leader. (BA-KP2)

While their discourse on the nature of the rank-and-file soldiers showed similarities with the intellectuals, the source of the problem and also its solution was interpreted rather differently. For the tacticians, the rank-and-file soldiers were illiterate and therefore followed anyone. What they lacked compared to other militaries was proper military accouterments, which would make them adhere to regulations and respect the chain of command instead of striving for positions and fame or listening to the wrong people:

> Our soldiers [in the resistance] were different from others' soldiers. Soldiers in our country were free and independent but rather undisciplined. Khmer soldiers were independent, but they only thought about gaining positions, fame and winning elections. They were not disciplined. Youn soldiers by contrast were disciplined and independent. Both Youn and Thai soldiers were all very afraid of their [respected their] leaders. (BA-KP2)

The illiterate rank-and-file soldiers of the resistance obeyed their commanders without wasting another thought, but military discipline was nevertheless missing. Loyalty that involves being over-credulous does not mean that it is correctly guided; thus, it can diverge from military discipline. Without proper military management, illiterate rank-and-file soldiers run the risk of obeying the wrong people. Lacking military education and discipline, they may obey just about anyone who gives them instructions or orders instead of adhering to proper formal military codes and regulations. In the end, there were two main sources of misguided loyalty that prevented adherence to military regulations:

> There were basically two sources. The first one was because they got drunk. The second was because there was a spy hidden in order to break the unity of the group. Once the spy said he did not want to go to fight because he was afraid of dying, they rest would hear and got scared. (BA-KP2)

Hence, since illiterate soldiers did not waste many thoughts about the veracity of people and their orders, they needed to be managed and monitored constantly. One way to do so was by keeping them busy all the time: "They didn't care much about anything. They only focused on following orders and playing or exercising during their break time. We had to allocate time for them to play, exercise, and train" (BA-F2). Constant control via military discipline and enough spare time to play games prevented them from clinging to spies or alcohol rather than the military chain of command.

Power Practices

Due to their own loyalty to one top commander, no mid-range commanders dared to speak about those commanders and how they treated their subordinates. Only seldom did they mention how they, as mid-range commanders, were punished if they did something wrong. And this, of course, was when acting in a way that went against 'military discipline', which would lead to 'classical' military punishments: "In case we did anything against military discipline, we would be punished to roll on the ground and things like that" (BA-KP2). And: "They taught us about military discipline (2) sleeping, eating, walking, and sitting. These four points all had to be clear" (BA-KP2). However, for the most part, the answer to questions such as "What happened if someone did something wrong?" was something like "That did not happen" or "We all loved each other". Given the fact that they all were long-time clients who served particular patrons until and beyond the day of the interview, whom they would never depict in a bad light, this answer tells a lot about their relationship to the patrons and is quite expected.

For most of the commanders interviewed, talking about their own leadership practices and their subordinates was much easier. At the center of their power practices were investigation and plan-making for combat operations as means of implementing their superiors' war strategies. This is what they stressed as their most important task. Hence, investigations and planning on a sand-filled table were a major part of the beginning of each mission to the interior:

> Before going to the battle, we had to assign one group to investigate the site. Then, we constructed a site at our place that resembled our enemy's position. We built the enemy's site on a sand-filled table, and then we reconstructed the situation and the geographic conditions. Then, we told our soldiers that our enemy's site was here, that we could walk from here to there, and fight from here to there. We meticulously planned all the steps and moves in advance. (BA-KP2)

While planning was the most important practice for the mid-range operator, military drill and training was most important for the rank-and-file soldier, who had to implement the plans later on. But sometimes the rank-and-file soldiers ran the risk of destroying the movement's reputation if they were unable to follow military orders and discipline, which is why a solution had to be found to solve the problem emerging in practice. For example, if a soldier raped a girl, the general understanding was that such behavior was beyond the tacticians' control. Thus, all that the military could do was to try to influence the surrounding situational factors positively and thereby solve possible discontents:

> Raping was beyond our command and it was very hard to prevent such thing from happening. But if we killed the one who raped a girl, we would lose a soldier. At least they had been with us for a while, so all we could do was to beg the girl's side [family]. Then the girl's side would agree and they could live together. So we could not always kill offenders without engaging in proper consideration. (AB-F)

Training and Punishment

The tacticians were those who implemented the training programs for the old military elite, the intellectuals, and others. While training was seldom conducted in units below strongmen, they were the ones who did the physical drill for the rank-and-file soldiers. A drill was a training exercise that was rather simple but regarded most decisive for surviving battles. Only a trained and skillful soldier was able to survive, which again could be compared to a boxing match:

> If they received proper training, they would be able to become skillful. If we provided them with no trainings at all, their presence in the battle was simply in vain. Just like when we go for boxing. If we don't know the techniques, we would make an unprofessional move. (BA-F2)

It is training in techniques that help soldiers survive, not religious instruments. Therefore, most tacticians stressed that while the use of herbal medicine may have been quite useful in normal situations, "in the forest", only proper skills in hiding and the skilled application of trained techniques helped a soldier survive a battle situation (BA-KP2). When asked whether monks helped in any respect, most commanders mentioned their practical assistance in conducting investigations and supplying their troops, not their religious influence: "Monks are quite good for providing our troops with some rice and information on the enemy" (BA-KP2).

Training was seen as essential because it awakened the soldier in the recruit:

> It is like pouring water on [recruits] to make them alive [...]. As I already told you, none of our soldiers was already born with military knowledge. Some even did not know how to use a gun, which is why they had to go through a lot of trainings. Soldiers were different from civilians. They had to live in collectives, wear the same uniform, and follow the same rules and regulations. Those who merely wore a uniform and carried guns around without having gone through military training schools were not suitable for being a real soldier. They knew nothing about the military, and only looked cool but ran away when battle came. Sending them to battle without proper training would mean to send them for certain death. [...] It is like you learn English nowadays. We are not born with knowledge in English. We have to learn it. Same goes for fighting. (BA-F4)

Providing recruits with political training was more or less useless because most combatants were illiterate. Basic awareness of political and strategic directions as provided during the roll call in the morning were viewed as more than enough:

> There was little political training. We only educated them to be aware of our strategic direction. Soldiers also looked at our political stance. We could not demand them to be highly-educated because how we recruited soldiers back then was different from how we would do it today. Many soldiers at the border were actually illiterate. So knowing military techniques, how to use the gun, self-protection strategies were the essential assets they needed to know. (BA-F2)

Since the illiterates were over-credulous, they always ran the risk of obeying misguided orders. In case they fell under the influence of poor leaders, education in military discipline and regulations was seen as a first step for setting them back on the correct path. If that failed, they were simply executed: "First, we educated them [in military discipline]. Second, we educated them. But if they still did not listen, they were shot" (BA-KP2). Since over-credulous illiterates risked spreading their misconduct and affecting the truly obedient soldiers among them, the leadership determined it was worth it to search for troublemakers for months on end. For the most part, they could be found, which certainly pleased the obedient soldiers:

> In case they joined our forces and later on created some sort of trouble, we told them we would kill them. For example, someone had stolen a gun and ran into a village to hide. We would definitely look for him and kill him. If we failed to arrest him this month, we would try again the following month. Sometimes they could escape because they were helped by their friends who were also soldiers. But one day when they went anywhere remote and reports came in that they did something somewhere, we would go and kill them. We did so to make our soldiers here [the obedient ones] cheerful. (BA-F2)

Seeking a proper investigation into their crimes was impossible for guerrilla forces, who had no court securing "human rights" as might be the case for "urban soldiers":

> I don't know for urban soldiers because they have their own court. But for guerrilla forces, if soldiers were not disciplined, we would simply shoot them dead. Then the rest would feel threatened. We did not care about human rights. (BA-KP2)

Sharing and Control
Sharing had to be generous and signal that the commander was willing to make a sacrifice in front of his soldiers. He had to demonstrate that he put the military collective above himself:

> I was quite different from others. Unlike those who always took some of the collective money for their own family, I always put all the money on the table. This particular group got this amount of money, so on and so forth. They could return as much money to me as they wish based on how much they love me. What is different? You love me! Others reserved money for five people, but they took the money home for themselves. But me, I gave them all the money. And then they would give me a certain amount of money back, depending on how much they love me. (BA-KP2)

In such a situation, the subordinates would certainly not dare to give their commander nothing, and they tried to show as much love as possible. In fact, it is quite likely that the subordinates had to compete with others in showing how much they actually loved him and so sacrificed from their own rations in return.

Sometimes leadership also entailed tricking people and pretending to be selfless although being offered something rather distasteful:

> I can talk a bit about Zhou Enlai and Deng Xiao Ping as a historical example. Soldiers got a bowl of rabbit soup for him, but he asked if the other soldiers also had it. When the soldier said no, Zhou asked the soldier to give it to the other soldiers. This was called killing the mind of the people. Leadership had to include this little trick. (AB-F)

Many commanders felt that they had to establish a certain distance between themselves and their subordinates in order to make themselves valuable. Although there may be a natural difference in brain capacity, not everyone felt that the difference in rank was always properly maintained:

> As a commander, we had to make ourselves valuable. When we slept, we slept on our own. It was the same when we ate. We shouldn't play around with the soldiers. We only talked to them if it was something related to our work. Once we finished discussing the matter, we must stop talking with them. This gave us value. But if we played around with them when we got drunk, we would become really cheap. (BA-F4)

Mingling with the subordinates was not an option for many commanders, as they felt it would put their status and value in question. In contrast to others, these commanders believed that they should not play around with them, drink with them, or sleep and eat at the same place as them. Rank promotion for the illiterates also only made sense in certain situations. Some were rather good in certain respects, but in the end, the commanders concluded that their brain capacity was simply too limited:

> Some Khmer people during the resistance period were illiterate but were able to analyze events more critically than the literates. So we could not ignore them. Sometimes they went to the villages and managed to convince several people to join our movement. We could provide them trainings and gave them positions limitedly. We could not give an illiterate too high of a position because, I am sorry to say, their brain could not function like a computer. They could command just twenty to thirty soldiers, but they could not command several thousands of people. No one could remember everything. In the history, very few of the illiterates could perform their job well. (AB-F)

Since the majority of rank-and-file soldiers followed those who provided for and treated them the best and lacked proper military education as well as the respective brain capacity, constant surveillance was the key to success. Like in a large family, everyone had to be monitored closely:

> I may compare it to a family. As a parent of seven children, we know which one is brave and which one is weak. A commander also knew his soldiers well. We knew who was afraid of battles. For example, we told them that we would need

to go to a particular place. The one who is afraid would convince us to stay, while the brave one would convince us to go. (BA-F4)

As in a family, knowing the strengths and weaknesses of your subordinates is important if one wants to assign them certain tasks. Knowing who was afraid and who was simply hungry was decisive, while proper training helped to solve the problem and form willing soldiers:

> Of course everyone was afraid. What people had in common was first hunger and second fright. Everyone was afraid but at different levels. We assigned people we knew. Those who were afraid would be assigned accordingly. We could ask them to cook. First, we trained them. Normally, no one was born with fighting strategies in mind, and no one was born knowing everything without learning. A yet-to-be designed piece of wood is not as attractive as its final products of furniture. The price is also different. We could only assign people tasks effectively if we knew them. (AB-F)

In order to make sure that they actually fought, personal surveillance was the key, since military leaders concluded that they could never trust soldiers: "I always stayed with them, so that they could not escape" (BA-F4). While close surveillance was essential, it needed to be followed by proper post-combat evaluation. Commanders and peers evaluated each subordinate unit's performance and made possible suggestions for improvement: "We evaluated our performance after combat operations. We could say which unit fought too slowly, which one too quickly. Then we thought of how to change that in future" (BA-KP2).

Tricking People by Creating 'Events'
After considering all emerging factors, commanders often drafted plans that included tricks that could be decisive for winning future battles. As one 'advisor' to a brigade commander put it: "If you ask me where I learned this from, I will just say that I simply sat down and thought. Problems would come up and make me think" (AB-F). Tricking simply meant to create events: "Some of these events happened by themselves and some had to be made up" (AB-F). He gave three examples of instances in which commanders created events: recruiting former Khmer Rouge soldiers, tricking the populace, and tricking the enemy. To understand how this tactic worked when trying to recruit former Khmer Rouge, his own biography was quite helpful:

> We could make up an event to make them believe in us. I am gonna talk about what we did in the past. No one wanted to be a Khmer Rouge, but for me I had no choice. Please don't say this in Pailin. I had my networks and I knew people. This person was a volunteer and that person was a volunteer [Khmer Rouge cadre]. While walking, we, pretending to have relatives as a policeman, pretended to drop a document and asked the person when he had joined the Khmer Rouge. After hearing this, that person would be very shocked. This was called making up an event. (AB-F)

Dropping a document that allegedly stated their background forced people to join due to the fear of retaliation by former victims in the interior. Creating events, such as a faked letter, also worked for gaining the support of the populace:

> This is how the event was made up: They were constructing a bridge. This village did not have a bridge, so soldiers made one for them. After the construction was complete, [we told them that] the Red Thai burnt it down. All these tricks were what we could never think of. I was the one who constructed the bridge, and I also was the one to destroy it. (AB-F)

And, finally, the enemy was tricked as well:

> During battles, we made up all kinds of events. If you look at [commander's name], he had just a few people but he managed to pretend [that he had big forces by igniting] hundred to two hundred burning bushes, thereby instilling fear among his enemies. When the enemy saw all these burning bushes, they thought that [commander's name] must have a lot of soldiers. (AB-F)

Battle-Hardened Roughnecks

In contrast to the tacticians from the patrimonial network of the leadership, the 'battle-hardened roughnecks' were the ones who made a career within the resistance. They were either promoted below a strongman favoring 'strong warriors', established a career as young recruits within the Cadet School, or defected from the Khmer Rouge after making a career there as former 'blank-page' leaders. Additionally, all have in common that they joined armed groups at a very young age. They were not mere operators but, rather, describe themselves as self-governed specialists in surviving battles from which they always expected to be "carried back in a coffin" by their subordinates with whom they fought in the front line (RC-F3). They acquired symbolic resources as strong warriors, earning them rank promotions below the strongmen or within the Khmer Rouge, which could subsequently be transferred to a mid-range position within the non-Communist factions. As former 'ordinary' soldiers, they interpreted their leading position as a collection of experience over the years, which brought them into the position of being their subordinates' 'father figure' due to their acquired experience and age, thereby having a similar habitus formation like the strongmen within the top command. However, they framed their rise to a mid-range commander not in terms of spiritual powers and fate but, rather, as being based upon a vivid and impulsive brutishness and fearless behavior. Rather than thinking first, they acted. Rather than being humble towards their soldiers, as emphasized by the strongmen, they scolded them. They scolded them like, as they would say, a father would. Similar to the strongmen, their resources were highly field-bound and volatile.

Habitus Formation and Lifecourse

Members of this habitus group established a career within armed groups at an early age. When they joined the Lon Nol military, the Khmer Rouge, or the resistance, they were still children, recruited at the age of fourteen to sixteen. Asking them about their childhood proved rather difficult, for the most part. Most responded that they did not remember much. Others did not remember much of their family life and at best answered rather vaguely that they were raised "according to our family's living conditions" (BA-F1) and that their parents were "very poor" (CC-F1, BA-KP3). One child soldier who went to a refugee camp at a very early age only replied after some thought: "(4) In 1979, we relied on humanitarian organizations. When the Youn came, I went directly to the camps. I never saw a Youn's face. (3) We were supported by UNBRO" (BA-KP3). All of the respondents were displaced and adapted to a life within the lower ranks of armed groups at an early age. Most felt rather abused as a subordinate whom commanders treated as they wished. They followed their leaders but did not receive any honesty in return: "I was like a dog, who was always honest with his owner. But they have never been honest with me" (RC-F3).

However, some of the mid-range commanders were also former Khmer Rouge 'blank pages', that is, child soldiers who made a career within the Khmer Rouge military due to their 'revolutionary class background' and 'purity' in age. They were drivers (CC-F2), mobile unit leaders (BA-F1), or mid-range commanders (RC-F3). Being fed up with the Khmer Rouge's policies or simply being on the list for purges, they defected to the non-Communists. Many reported that they initially joined the Khmer Rouge to bring back the deposed king during the early 1970s. Following the king remained a recurrent theme, as many of them stated that they joined Funcinpec for the same reason (CC-F2; BA-F1; RC-F3). For the sake of coherence, they did not join the Khmer Rouge due to being particularly Socialist but, rather, due to their loyalty to the king, which may have simply been a means for justifying their engagement with the Khmer Rouge or truly the main motive behind their actions. In the end, most roughnecks did not defect right after 1979 when the time was ripe but waited until they were threatened to be purged by the party center. Only then did they rediscover their loyalty to the royalty, taking hundreds of men with them (BA-F1; RC-F3). In the end, this might also be a reason why all of these former Khmer Rouge interviewed happened to be members of Funcinpec (ANS) and not KPNLF.

Access to the Field

Practically none of the roughnecks had prior contacts in the field. All of them were recruited at an early age, enlisted in a camp as a child after they became too old for the UNBRO distribution system, which excluded grown-up males, or even may simply had a commander as their neighbor within the encampment who recruited and promoted them on their way (BA-KP3). Access to the field, however, was a bit trickier for the former Khmer Rouge. Not everyone greeted them with open arms although they were oftentimes welcomed as a useful and battle-hardened fighting force. Within ANS, a unit called the 'Black Eagle Unit' was regarded as a pool for collecting the incoming former Khmer Rouge cadres.

Similar to the strongmen, because they came from a lower illiterate peasantry and had no significant social resources within the field, battle experience was the roughnecks' main symbolic resource. In contrast to the strongmen, the roughnecks barely experienced a life outside of armed groups. As child soldiers, they adapted to an 'unethical extremism', in which violence becomes legitimized because it is interpreted as a 'necessity' of soldierly life. Ethics only hamper their military performance and put the lives of those who believe in them at risk. The roughnecks' main symbolic resource was their practical experience and, similar to the strongmen, survival of battles that gave them an increase in status within the field.

Classificatory Discourse

Really good commanders were those who came from below, that is, those who were formerly rank-and-file warriors who went through the 'school of hard knocks' in battle:

> Some people became commanders without having previously been an ordinary soldier, which is why they were not aware of real leadership skills. By contrast, those who were rank-and-file soldiers before and subsequently became commanders were usually very good. (BA-F1)

Having gone through many battles made a soldier a really tough warrior:

> Fighting in battle was basically the same like in trainings. I was involved in so many battles in 1975 that I wasn't afraid of anything anymore since I saw it all before. As long as you did not experience combats before, you would run away just at hearing the sound of a thunder. When I arrived at the camps, I saw how many people did not even know how to unlock a gun. For me, during the Lon Nol regime, I used to fight in a battle lasting continuously for forty-two days. That was how I trained myself. My experience made me brave. (BA-F1)

All respondents from this group highlight how they rose as tough warriors despite their lower peasantry background. Another respondent tried to formulate his pride as modestly as possible: "That sounds like I am showing off, I know. I only received little education, but I was a capable fighter!" He continued, "I myself used to fight in countless battles. I knew all the techniques. I was not afraid of anything. I was strong – both militarily and verbally" (RC-F3). And, last but not least, a company commander explained:

> Let me tell you one thing. I am not saying that I was involved with hundreds or thousands of battles. I am only saying that I got involved in countless of them, and this made me know all the techniques. [...] You may not believe what I

am saying but during my fifteen years in combat, I never lowered myself to the ground. During combat, I always walked and ran around at the front line to check up on my soldiers and motivate them to fight. (CC-F2)

Having gone through the tough school of battle experience also meant that you sometimes needed to be unethical in your leadership style. You needed to be tough and rather brutal. Violence was a necessary part of being a soldier. Scolding people was part of leading well, and ethical rules as set by a *kru khmai* were not very helpful:

My *kru* always told us not to scold others; second, not to touch other people's wives; third, to respect elders; fourth, not to burn other people's houses. However, as a commander, if we spoke nicely to soldiers, they wouldn't follow. We had to use violent means to command them. Using morality would simply fail. (RC-F3)

In general, monks were not believed to be very useful. Battle made the roughnecks tough and taught them that there were neither sins nor any god: "I don't believe in god or sins. If they existed, they would have appeared and saved all those kids and people who were being killed by the Khmer Rouge" (RC-F3). Even worse, he went on to say that people were afraid follow the spiritual guidance and would end up dead while hiding at the back:

R: Those who were afraid would follow the *kru*'s advice. Those who were not, wouldn't.

I: So who was safer then?

R: From my observation, those who were afraid and stayed at the back usually died. (RC-F3)

Anti-Intellectualism
In contrast to the tacticians, roughnecks did not think first; they acted. They did not talk much but ensured that a job was done: "I was a serious person. I didn't say much but would rather do more. For example, once I was committed to occupy an area, I would ensure that it would succeed" (RC-F3). Instead of wasting time, they focused on the essentials: "I did things I knew I needed to. And ignored, what I had to ignore" (CC-F1).

They constantly complained about intellectuals and people who believed that they were above others and demanded everything for themselves while others had to suffer. One commander complained about one of his colleagues:

Some people may not have received as much education as him, but still they came to join the resistance. They even dared to sacrifice themselves for the country. He who could always sleep comfortably never considered that. We were even cheaper than a dog for him. At least dogs would be fed by their owner. (RC-F3)

Unlike the tacticians, who only implemented commands, they were uncontested in leading *their* men:

> At first, for example, I commanded about one hundred soldiers, so all hundred soldiers listened only to me. Besides me, they didn't listen to anyone else, not even to the commanders coming from heaven [that is, those believing in their absolute superiority]. If any big commander came and commanded them, they would not go, and they might even shoot him dead. At that time, we shared soldiers, so leadership also had to be shared. (CC-F2)

Becoming Strong: Adaption to the Role Model

In addition to actual combat, another key to success centered on subordinates imitating and adapting to their commander's behavior, including his style of walking and commanding. By imitating the commander, a soldier could learn how to look tough and speak accordingly:

> I also learned from my own commander. Even the way how he walked already made soldiers afraid. My commander taught me and told me the reason why he kept on scolding me. It was because he wanted me to be good and once he dies, I would be his successor. I never imagined to be in such a high position back then or something else like this. I learned a lot from him. (BA-F1)

Imitation and scolding made soldiers tough and strong enough for a leadership position. The strength of a soldier depended on the strength of his commander, who was like a father to him:

> A soldier could only become skillful because of his commander by seeing his example. The commander couldn't stay at the back after having commanded to start an attack [thereby leaving the soldier without a good example]. How good a child can be depends on the parents. If the child listens to his parents, he can be considered as a good child. But if he doesn't, he is bad. Let me put it this way. (CC-F1)

In the end, soldiers were willing to fight but needed good guidance by a tough warrior who already went through all the battle experience when he was younger. Because most of them were of 'hot-blooded age', soldiers were willing to fight: "Soldiers were eager to fight because they were in their hot-blooded age of 25 or 26" (BA-F1).

However, all respondents stressed how important it was to be scolded and to scold others. By scolding them, their 'hot blood' could be tempered and used for combat operations without them losing their mind:

> Normally, we don't want others to scold us. But for soldiers, they needed to be trained so hard that they were no longer angry when someone scolded them. Soldiers had a gun in their hands at that time. Hence, they could shoot anyone at any time if they were angry. But when they had been trained hard, they were no longer angry. (CC-F2)

All comrades were buddies, though the commander was the strongest buddy among them all. All soldiers were seen as equal, but he was situated a little bit above them.

Hierarchies among Comrades and Friends: Being Equal but Stronger
Similar to the strongmen, the roughnecks characterized their leadership style as much like parenting, in which they "care about soldier's problems such as whether or not the soldiers or their family are starving" (BA-KP3). Using the word "parent", which is also used to denote a boss in the Khmer language, one commander remarked: "Normally, soldiers wanted to see their commander stay with them when they are in trouble. When they saw their parents [commanders], they would be happy and brave. I have learned a lot in my fifteen years as a commander" (BA-F1). The commander had to stay with his 'children' and be a role model for them. He had to lower himself to their position and make them happy by spending time with them, playing soccer with them, sharing some drinks with them (BA-F1). The 'children' in turn imitate his behavior, just as the commander did with his own commander.

A commander's behavior even verged on friendship, in which soldiers could even use the rather impolite Khmer prefix "a-" in front of the commander's surname to denote close friendship. Most soldiers would not dare to use this informal form, which is why the commander with a modern style leadership did when talking to his soldiers:

> I led soldiers using modern leadership style, not the classical one. A commander who used the latter approach usually could not win the battle. I made friends with my soldiers. Everyone was my friend. If someone was younger than me and dared not call me 'a-', he had to call me 'commander'. But I had to call him 'a-'. I slept with them, and they slept with me. I was the same to them, and they were the same to me. We ate together. Sometimes I even cooked for them because we all were friends. We were willing to save each other during the battle. So why weren't we willing to make friends with one another? Even though we became friends, soldiers still listened to me. (CC-F2)

Equality in spite of the military hierarchy is what he called a modern leadership approach. Instead of insisting on strict hierarchies, the roughneck commander almost enforced equality during casual conversations and when giving orders:

> I commanded them like that: "Hey friend, you have to bring a group of soldiers to ambush Youn soldiers along their way. No matter how long it would take, you have to attack them successfully, OK, friend? You have to set up mines to kill them". When he said OK, he would do it, I would mobilize some forces. We just told them what to do. Soldiers felt warm because their commander was their friend. They were afraid of me during the battle. But after the battle, they no longer were. Sometimes they even poured beer onto my head, and I was fine with it. As a friend, they could do whatever they wanted to. Sometimes when there was nothing to eat, they scolded like, "You mother fucker, I couldn't find anything to eat!" And I replied, "You mother fucker, I have nothing to eat either". (CC-F2)

However, although they could be considered friends based on their linguistic practice, his soldiers were still afraid of him during battles. Even though he was a rather 'chummy' parent, the commander was still the parent who was respected and even feared by his children, and not only seen as a friend. Superior strength and experience made the parent a respected:

> Once they decided to follow me, they must stick with my commands. I am not embracing communism [he means Khmer Rouge principle of Democratic Centralism], but as a commander, we must always be stronger than our soldiers. Otherwise, we couldn't command them. (RC-F3)

What characterized the difference between an ordinary soldier and his commander is that they were equal in nature but the commander was stronger.

Power Practices

An essential part of disciplining soldiers to become good soldiers was not to train them but to make them strong through experience in battle. Thereby they learned when tactical retreat was the best option and when engaging in fighting would be best. Military training and its application in battle was highly risky:

> They were barely useful in combat. Once the bullet approaches soldiers, they forget all the things they have learned before. However, they had to receive training because military rule required them to do so. If they applied what they have learned during actual shootings, they would die easily. (CC-F2)

The roughnecks refer to the very same knife that training sharpened for the tacticians. Only confrontation with battle sharpened it and made anxious soldiers brave:

> I: How could those who were afraid during battles be helped?
>
> R: If we knew that a knife was not sharp enough, we brought him close to us. When we went to a battle, we brought him along. We threatened and encouraged him, or even held his hands throughout the whole combat operation. Of course, he would constantly say that he is too afraid, but we would also go on saying he shall not be afraid and motivate him to overcome his fears. Next time, he wouldn't run away anymore but be brave like everybody else. (BA-F1)

Hardly any of the roughnecks used magical items to make people brave (BA-KP3). Only a few believed "in some magic":

> I also believed in some magic. My *kru* told me that I wouldn't die as long as I do not lower myself to the ground, otherwise I would get killed. I followed his advice once, and then a second, a third and a few more times, and I was always safe. And when my soldiers saw their strong commander, they also became strong. (BA-F1)

Rewards and Punishments

Likewise, a central aspect of promotion is bravery. Bravery as observed during combat would be rewarded with promotions in rank: "Amongst our soldiers, we had to know which ones are strong and which ones are not. We had to observe them, and see who is brave in combat" (BA-F1). Providing food and taking care of his subordinates was the top commander's priority. He had to provide well for and ensure everyone's well-being. A company's commander praised his commander-in-chief for doing this job extremely well and even ensuring that his men had a balanced diet as well as medical precautions:

> He was a good commander because he always fed them well. He gave twenty chickens to each unit, so no one had to starve. We were not even allowed to eat prahok because if we ate energetic food, we could get malaria easily. We always had to drink hot water as well. If any unit did not boil the water, he would come and scold the unit. He bought us medicines, and assigned someone as a cook to boil the water for us. (CC-F1)

For the most part, however, the roughnecks did not waste much time before acting. As one commander explained, they act (brutally if needed) and do not speak much: "I didn't talk much to my soldiers" (RC-F3). Most highlighted that they had been rather brutal 'parents', not only scolding their soldiers all the time to make them tough but also punching them if necessary: "I could only speak nicely if I wasn't angry. As soon as I turned angry, I wouldn't talk more than three words. I would punch if I needed to. I never cared about committing sins at all" (RC-F3). When enraged by their soldiers' disobedience, the roughnecks would simply hit them: "If they didn't obey my orders, they rather shouldn't come too close to me; otherwise, I would drag the stick and hit them hard without wasting much thought about it" (RC-F3). Not only beatings could be used for physical correction and drill but also tortures, such as forcing a soldier to carry heavy items in the sun or stand within an iron cage: "If soldiers did not follow orders, they would be punished to stand carrying heavy objects under the heat of the sun or they would be put in a cage for a few hours under the sun" (CC-F2).

Not only did they not waste many thoughts and words before hitting people but they also hardly thought twice about killing wrongdoers after having already warned or threatened them: "If they still did not follow after being threatened before, one thing we could do was to shoot them dead" (BA-F1). Killing someone outside of battle was not just bad for the wrongdoer who was killed. Much more important, it constituted a highly dishonorable death for a proud warrior that worked a circuit:

> Having killed a few undisciplined soldiers would also threaten the rest. It was OK to die in battle, but dying because of being shot by the commander was different. Other soldiers would hear that and not dare to reject any orders. (BA-F1)

Blank-Page Operators

'Blank-page operators' within the Khmer Rouge were situated at the top of the command, even in positions as brigade commanders. Within other groups, they would have been regarded as leadership, but the organizational mode of the Khmer Rouge was different: because they were not in the position to set up rules in the field but only operated upon the rules set by the Central Committee and military directorate, they were treated as 'mere' operators in charge of implementing and adjusting commands. In short, their formal rank could not be compared to the volume of power that commanders at the same level had within other military forces. Furthermore, battalion commanders interviewed for this project had a similar habitus. The reason that the mid-range command in both groups was comparatively uniform lies in the strict screening for recruitment of commanders by the top command. Since they all had to meet two essential recruitment criteria, their backgrounds were very similar: All of the commanders interviewed were children when they joined the Khmer Rouge during the early 1970s (between thirteen and sixteen years of age), and all of them came from lower-class peasant families or something similar (one, for example, had a father who was a traditional music singer in the countryside).

Habitus Formation and Lifecourse

> I wasn't an adult yet, still a teen. (BA-KR1)

Most of the respondents within this group hardly remembered their childhood, and if they did, they connect it with a broken family and hardships. As a result, they did not talk much about it. Those being recruited at an age of thirteen or fourteen simply stressed how poor they were: "I was too young to know anything about my parents. They were farmers. Right now they are climbing on palm trees" (RC-KR4). Unlike him, most of the others did not see their parents again after joining the Khmer Rouge as children. All respondents stressed that they came from poor, highly impoverished peasant families. Many also highlighted sufferings connected with the early death of one parent – in most cases the father. Due to their living conditions and broken families, most respondents had to help with the farming or by earning some extra money in order to keep the family afloat. Each respondent was send to a pagoda after a while so that he would be able to go to school and simply due to the fact that he could live and be provided for as a monk. There they stayed as a monk novice for three to five years until the war broke out after Lon Nol deposed Sihanouk in 1970. Cadres interviewed for this project were largely recruited at an early age between thirteen and sixteen. At first they were put into youth organizations where they were indoctrinated, oftentimes starved and tortured as well, and their commitment and revolutionary spirit tested. Only after passing these commitment tests were they eligible to serve in a core organization, such as the military and security apparatus. During the initial phase, their spirit had to be adjusted to the ideal model as set by the leadership, and they were only chosen as 'absolutes' for the party's core once they proved compliance with the party line in thought and action. Most simply tried to imitate the language and habits of what they believed complied with the idea of what a good cadre was in the eyes of Angkar.

The coup against Sihanouk was cited as the main reason to join the Khmer Rouge by most of the respondents. As monk novices, they felt that the social and cosmic order was shaken by this act and had to be restored. Or, for some, it was a loyalty they felt to their ancestors who, in turn, owed their loyalty to the king:

> I started to run into the 'marquis' jungle following our king's call. We actually did not know what the 'marquis' is. We only heard of that and then followed each other. [...] Because a lot of our ancestors owed their loyalty to the King Father, we had to follow his call when he was ousted. (BA-KR1)

Not just individual monks but whole villages followed the King Father's call, according to one respondent, who stressed that back then the US supported Lon Nol's coup only so that they could come back today to help:

> The entire village joined. King Sihanouk told us to go into the marquis forest in order to survive because Lon Nol was staging a coup. Now the Americans are coming to help Khmer every day. As long as we help each other, we all will survive together. Before, Khmer wanted to gain power. But for me, I no longer want it. My education is also very low. (RC-KR4)

For the Khmer Rouge, only US imperialists were behind the coup in 1970. Most were attending school at that time and lived in a pagoda, but since war destroyed all of the infrastructure, some stressed that this pushed them into armed groups – either within the state or within the guerrilla, as they were forced to end their studies rapidly when schools were closed (RC-KR4). However, not everyone recited a master cleavage as their reason for joining. One respondent admitted that he joined "as requested by the king. But actually, I don't know why I did this" (BC-KR1). Since they were young, other motives they gave for joining included, for example, a fascination with being a soldier:

> The reason why I joined the military was because I wanted to carry a gun and fight. I didn't have the idea of loving the nation or hating anyone else yet. I was like a hot-blooded teen. I always prayed that I could go to fight because I really liked it. (BC-KR1)

It was neither a struggle to liberate something or someone, nor to end imperialism, feudalism, or capitalism. Within a society at war, being a soldier for a child had its very own appeal and may have promised a better, more exciting or simply safer life. Each of these children, however, had one main reason for their internal promotion to higher ranks: contacts to the upper echelon. Some even knew some of the Khmer Rouge leaders from early childhood onwards or had a relative within the leadership (e.g., RC-KR2; RC-KR4). Hence, although they had to fulfill strict criteria to join the Khmer Rouge forces and had been ideal 'blank pages' who proved willing to serve Angkar, they also had social resources facilitating their rise in rank up to brigade, regiment, or battalion commanders. For some, this may have been the main factor that differentiated them from other 'blank pages' who did not establish a career within the military forces. In addition to the attacks against the family as an institution, the party's central command proved keen to patrimonialism as well. While lower Khmer Rouge ranks were introduced to the military by other cadres as well, they lacked valuable social resources that would enable their 'constant rise' to positions right below the top command. Here, family-based clientelism was oftentimes at work, as many positions in the military below a patron were given to family members in order secure an entrusted grip on the apparatus.

After the Vietnamese invasion, these networks reconstituted themselves relatively quickly. But not everyone followed willingly:

> At that time, me and other soldiers really did not want to continue being soldiers because we gained nothing from going into war. There were only death, injury, disability, and separation from our wives and children. Since we received this political influence, we had to go on. In my case, I was with the Khmer Rouge, so I only got political teaching from the Khmer Rouge. They told us that if we went back to the country or allied with the State of Cambodia, we would be killed. So we came up with the thought that if we went back, we had no choice but to be killed. However, if we stayed, either we died or we survived. Actually, if we were asked whether or not we wanted to fight or go into war, we would of course say 'no'. Yet, I was politically influenced by the Khmer Rouge. The way they influenced me was through talking about 'death' all the time. (BA-KR1)

Although many were fed up with the constant internal purges and threat of imprisonment, they stayed within the forces until the late 1990s when they saw a chance to change sides. Constant threats making it clear that it was better to stay made them do just that: "The government said that as soon as we were arrested, they would cut open our belly" (RC-KR5). Staying at the border and resuming fighting was the safest option for them. Especially during the mid-1990s, dissatisfaction was on the rise within the operators' ranks:

> We were not satisfied with our leaders like Ta Mok and Pol Pot because we knew they wouldn't easily give up. As soldiers fighting in battle, we knew how war destroyed all of the infrastructure. But still our leaders pushed us to resume fighting. (BC-KR1)

Their discourse, however, was shaped by what they understood as their main symbolic resource: practical experience and obedience to the top command (the 'collective'). They interpreted the Khmer Rouge discourse rather unintellectually as 'merely' practice guided and saw no connection to the intellectual part of the top command's discourse on 'normed practice'. As a result, they translated the principle of 'experience by normed practice' into 'learning by doing'.

Classificatory Discourse: From Norm to Practice

> The Khmer Rouge had that theory that if it was too difficult to provide trainings
> to soldiers, one should send them directly to combat. (BC-F4)

While it does not seem to be a theory of 'the' Khmer Rouge to send recruits
into battle immediately, the mid-range of the Khmer Rouge understood the top
command's directives to imply just that. Much of the intellectualized theoretical
formation was lost in translation within the operators' ranks. These blank-page
leaders adopted the leadership's discourse not like blank pages but in rather
unintellectual terms as a primacy of what they believed to dispose of most:
practical experience, obedience to the collective, and willingness (as measured
visible 'activeness' and unquestioned commitment) in executing commands. As
a result, they mirrored many of the top command's concepts but slightly adjusted
them according to their own habitus formation.

The Good and the Bad by Command

For the Khmer Rouge operators, combining theory as learned in some political
indoctrination sessions with practice was simply too complicated to manage:

> It was difficult to practice and learn at the same time. Sometimes we still fell in
> danger even if we had learned the theories because we were not very sure with
> those theories. They could also gain good experience quickly when they could
> fight directly in the battle. (RC-KR4)

As a result, pure practice – instead of formulaic and intellectualized practice – was
enough to make soldiers good and brave. This was one of the major adjustments
within the mid-range's implementation of the anti-intellectual intellectuals'
directives and power practices. Putting them right into the contested area, i.e., the
battlefield, was seen as the best way to form good soldiers, not simulating combat
and following formulaic provisions aimed at setting up proper 'revolutionary'
practice. It was so easy that even the researcher and his assistant were invited to
learn how to 'practice' right away:

> There was training right during the battle and in midst of the contested area.
> This was what made them become good soldiers. You two [the researcher
> and his assistant] can also go with me now [to learn how to fight by practical
> experience]. (RC-KR4)

Fighting was easy and could be learned in actual combat. The Khmer Rouge
operators were truly un- and anti-intellectual. For them, people started their lives
as blank slates. Becoming skillful and obedient was, therefore, a twofold process
of gaining on-the-spot experience and constantly following the command. The
soldier was viewed as the product of the commander – how he commanded him and
which ideology he poured into him. Soldiers simply mirrored their commander:
"The ideology spread to us could let us do anything. If the ideology made us bad,

we would become bad. If it made us shoot, we would always shoot" (RC-KR4). Therefore, the blank-page leaders also preferred to recruit children. This was because children would become real tough fighters and could pose a problem to each enemy military they encountered. They "were born with strength" and could be instrumented accordingly (RC-KP1). While talking about his child recruits from the years of civil war, one commander highlighted that they even prove brave and skillful today, twenty years later, when fighting for the Cambodian government against the Thai in 2011:

> Sure, even soldiers with this height [pointing to a rather low level] carried a gun to fight. During the battle with Thai soldiers at Phnom Trop [just recently his former soldiers fought in the standoff with the Thai over Preah Vihear], those short soldiers never ran away from battle. Their height did not meet international standards back then. Some were just about 15 or 16 years old. But the Thais could not easily defeat us over here. (BC-KR1)

The good thing about children was that they adapted easily to a life as a soldier. The same applied to the blank-page leaders themselves: They were also tough fighters due to their practical experience and obedience to the collective. Therefore, they did not spare lengthy talks about their own involvement in battles or recount how many scars they had due to bullets or how many times they were wounded in battle (e.g., RC-KR2).

In the end, combat as such was easy anyway. Even for the commander, planning combat was a simple part of the daily routine. Asked about what was necessary for a soldier to become a good commander, the reply of one respondent was, therefore:

> Are you asking this question because you want to become a military commander? How do you organize each combat? It was just like when you go to eat something. We just needed to know how many plates and how many guests there are. It is nothing hard to do. (RC-KR4)

Democratic Centralism to Fight Self-Interest
Following Khmer Rouge discourse, however, a soldier is the product not just of practice but also of his commander's commands and ideological stance, which becomes inscribed in his behavior. The main principle for determining whether a soldier or a commander is good or bad is whether or not he acts in his own self-interest. Individualism and self-interest were to be prevented by sticking to command and to the principle of "democracy" as in democratic centralism: "It is said that in democracy, one shouldn't do anything on your own. There should always be discussion with others" (RC-KR1.) Therefore, nobody should act on his own without the approval of the collective, whose will is handed down from the center by the respective unit's superior: "Individualism was not allowed. But when we have democracy, everything is possible. During battles, if a commander does not give any orders, soldiers cannot operate on their own" (RC-KR1). Since the superior is viewed as more knowledgeable, soldiers have to follow his decisions.

The soldier himself does not know much and even should not know more than necessary:

> We had to listen to both: ourselves and our commander. We couldn't do things as based on only ourselves. We would be wrong if we tried to claim that we were the most outstanding and knowledgeable soldier. We shouldn't know too much. We had to listen to our commander's commands. We shouldn't do work that was not commanded of us. (RC-KR2)

Within the discourse of democratic centralism, the individual has to accept the wisdom of the collective. In the end, the soldier only executes what the commander orders and is relieved of his personal responsibilities when executing the commands given. He is responsible not for the orders but for properly organizing their implementation. The soldier can discuss with his peers, but he has to stick to the command.

The collective is omniscient, and soldiers are the ones who have to sacrifice their own lives for it: "Everyone spends money, but soldiers spend their own life" (RC-KR3). A lack of sacrifice for the collective causes the collective goals to fail. Therefore, group conformity is most the decisive factor for the collective, even if it means that the individual group members have to sacrifice their own lives. What needs to be prevented is exertion of too much freedom by teaching 'democracy': "At first, we taught them about democracy. They should not be too free. There must be rules and regulations. Later, they could fight, turn skillful and become commanders" (RC-KR1). A purely 'collective spirit' stands at the beginning of everything. Therefore, constant surveillance and instruction are necessary: "We had to make sure that there was not too much freedom given to the soldiers. We had to educate them" (RC-KR1). A good commander, as a good soldier, does not think about himself but has to know his own weaknesses and be aware of the limitations to his perception due to his position in the command. He has to know himself and whether he is acting in a way that strengthens the collective clearly:

> A really successful commander has to know himself, what he is doing, and how he is doing it clearly. He always has to know whether what he is doing may affect the country or the society or not. He can't just do things based on his personal ideology. (BC-KR1)

The willingness to serve the collective is decisive when it comes to the question of whether or not soldiers fight. It is a question of their willingness to sacrifice themselves because if a soldier lacks it, he would leave his weapon unused: "But the most important thing was their willingness. If they were not willing, that would not be possible. No matter how many weapons we had, once people did not go to fight, weapons would not be used" (RC-KR3). As a result, a soldier needs to be constantly rectified or 'alerted' as to what his duties are. Otherwise, he puts himself above the group and undermines its strength. His actions could be a possible sign of him being a counterrevolutionary spy; thus, they are clearly also an indication of the need for rectification and vigilance.

Constant Rectification and Vigilance

> They become skillful only when their commander was always with them, to constantly instruct them, just like me. We controlled seven hundred to eight hundred soldiers. One unit here and one more there – that's why we always needed to instruct them. We never missed any month of the year for instruction. (RC-KR1)

Constant instruction was needed to control the lower-ranked soldiers. Similar to the battle-hardened roughnecks' principle of scolding each and every one of their soldiers to make them tough fighters, many mid-range Khmer Rouge point towards the need to scold and constantly harass a recruit in order to make him a tough fighter. The soldiers needed to be rectified since they could not be as "straight" as their commander without having gone through all the experiences he did. However, some needed to be scolded, rectified, or, as one respondent put it, "straightened" a bit more than others:

> But as a commander, you had basically three types. First, some soldiers would obey even without being told to do so. Others wouldn't obey unless they had been scolded first. And some other wouldn't obey unless we had given them harsh punishments. Some soldiers were curved at the bottom and on top, but straight in the middle. They were rarely evenly straight like a ruler. When we commanded, if we couldn't understand their mentality, there would be an argument. For example, as a regiment or battalion commander, if we didn't say the right things [to the respective type of soldier], no one would listen to us during the meeting. (RC-KR2)

As the analogy went, some clay needed to be formed by scolding to make it straight. Collectives always encompassed various types of people who required differing degrees of rectification. The aim of rectification was that new recruits needed to strip off their individuality and follow regulations and rules. Before joining the military they acted 'freely', but they had to obey the rules once they enlisted. If not, they might be assassinated:

> Military service could change some of the recruits. For example, they were used to do everything freely before joining the military, but after joining they had to follow certain rules and regulations. And they had to be careful at any time because they, as fighters, might be assassinated at any time. (RC-KR1)

The blank-page leaders incorporated one major rule into their command: constant vigilance. Soldiers could be assassinated at any time if they did not follow the rules. They were either killed in combat because improper discipline warranted killing them if they implemented a command incorrectly, or they were killed by and after rectification. Strictly speaking, there was no specific skill that a soldier needed to demonstrate upon his recruitment (since he was viewed as a blank page anyway), but he had to adopt a 'concept' of being alert and vigilant all the time:

There was no such thing like skills during that time. If we talk about skills, we would have to talk about the Communist concept, which said that we had always had to be dressed even if it's one o'clock at night. We had to be ready, regardless of the time. After finishing our breakfast at five, we had to get ready for fighting. (RC-KR2)

Being ready to follow orders that were handed down by the chain of command was what characterized the skill of a combatant. Regardless of the time of day, he needed to be vigilant.

Command is Command and Those Who Do Not Follow Are Spies
Blank-page leaders constantly stressed that there was no way to circumvent obedience to command, since it was the essential skill of a soldier to serve and constantly be ready to sacrifice himself. Even when starving during battle and pressing on meant that some soldiers would die, Khmer Rouge were obedient to command: "During the battle, if we had been commanded to go forward, we had to follow, although we were starving. Sometimes during the battle, we had to separate. As we were starving, we dug for some cassava to eat" (RC-KR5). Though at times reluctantly, a good soldier always followed commands, even in cases where they saw that a command given would result in civilian losses. They were entitled to their opinions at some stages, but in the end, they had to follow:

I: Have you ever been reluctant to follow an order?

R: There were some cases. For example, civilians were living around the area being singled out for an attack. This made us reluctant to fight. If we attacked that area, we would harm these people. But we had to follow the command.

I: Couldn't you refuse in such cases?

R: In the military, a command is a command. We may raise something because objecting the command was impossible. We may raise some reasons to change the plan. If they accepted, it was fine. But if not, we would have to go on as commanded. (RC-KR3)

Each soldier needed to follow the command. If he did not do so willingly and with a proper 'political spirit', he would face troubles. Soldiers needed to do what they were told to: "When they were told to stop, they would stop. When they were told to fight, they would fight. If they didn't respect the command, they would be in trouble" (RC-KR2). Angkar as the omnipresent force was cited as the entity that determined the direction of the collective. The soldier was obliged to follow its words: "During the battle, no matter if we were at the front or at the back, we had to fulfill our obligation, following Angkar's direction. Let me just call it Angkar" (RC-KR5). Counterrevolutionary spies caused indiscipline and failures within the military, hiding within the units and causing all kinds of troubles. Those who did not prove willing to be rectified and cleansed were to be purged, since these "were spies hiding within our forces" (RC-KR1). They themselves believed to be the incorporation of the top command's ideal: formed by practice, loyal automata of Angkar, unconditionally vigilant, and vigorous in fulfilling their duties.

Power Practices

Being a spy represented the ultimate assertion of treason. Within the logic of democratic centralism, the plan was always assumed to be correct. The only problem that most cadres within the Khmer Rouge faced were spies who sabotaged the proper implementation of commands and tried to put the blame on the leadership. Therefore, my research assistant and I encountered problems talking to Khmer Rouge about their practices. Oftentimes, respondents suddenly accused us of being spies, who were trying to put the blame on them for the killings, starvation, and all kinds of failures under the Khmer Rouge. While asserting that I have other intentions than pure research, they also suspected that my assistant might be a Vietnamese (or 'Youn'):

> You [the researcher's assistant] must be a Youn as well. Are you Youn? Germans think that their soldiers are strong, but they also want to know how Khmer Rouge soldiers are like. Is that the truth [of your intentions to interview me]? (RC-KR4)

Oftentimes, other Khmer Rouge combatants sneaked around during the interviews. Some had served in the same unit as the interviewee; others had been their subordinates, and still others even their former commanders. This made it even more complicated for interviewees to talk at ease. Many tried to deny any kind of involvement in 'bad practices' such as violence or even killings. However, their attempts not to talk about it were seldom successful and resulted in some contradictions:

> Good leadership means to be gentle with honest people. If people were not honest, we couldn't be gentle with them. I didn't kill anyone. I only educated them. We told them not to do that again next time. It was not hard to educate good people. They were already good. In total twelve up to thirty people were good, and only a few people were bad. I told them that morality was their own business, not mine. I was only here for their death. I told them that if they didn't listen to me, they would face failure next time. We wouldn't accept if they raped people's daughters, shot anyone, or stole people's properties. If they did so, they would die one day. As a commander, I wouldn't wish that my soldiers would one day die. (RC-KR4)

They mostly referred to sufferings that were felt by a soldier when being commanded to do something he did not like to do. Although soldiers suffered a lot because of their proximity to death, they were simply instruments of the government or the respective leadership:

> Now let me ask you two [the researcher and his assistant] a question. Before people die, do they suffer from convulsion? This is the main point of the story. Before dying, people convulse. No matter what, before dying, do they convulse? Political affairs are a government's task because we are their instruments. Do you understand? (RC-KR4)

Although being recruited as children and constantly indoctrinated thereafter, they nevertheless incorporated a different understanding of the Khmer Rouge discourse and accordingly opted for very different practices at times. While many of the practices from the top command were implemented in a tight-knit system of control, many of them were more or less slightly adjusted.

Transmission I: Recruitment and Promotion
While biographical questionnaires were used for recruitment in some cases, most units did not rely on a formalized procedure. For them, in order to make sure that the background of a recruit was 'clean' and revolutionary-minded, they had to check up on two factors: his background and previous activities. 'Background' here means coming from the peasantry in a sometimes rather broad sense, while 'previous activities' was oftentimes interpreted very individually and at most refers to whether the recruit had been a thief or something else that accounted for morally 'bad' behavior in their view:

> We only checked their background and previous activities. If someone didn't have a good background or used to be thief [bad activity], we wouldn't recruit him. We only had to be very careful with this. We only had to know their background, like whether or not they used to be a thief. (BC-KR1)

As a result, the commander concludes: "Their lives were suitable for becoming a soldier within the resistance movement. They didn't do anything bad" (BC-KR1). It was thus easiest to rely on children, since they most likely had not yet engaged in 'bad activities'. Nevertheless, keeping a watchful eye on the background of a recruit was not enough. After recruitment, they had to remain vigilant about the recruit's actions. The mid-range command sensed an omnipresent threat of spies coming to sabotage the military. These would have to be identified and subsequently rectified:

> R: We always had to be careful with newly recruited soldiers. Sometimes our enemy assigned someone to join us. Sometimes they came to kill our soldiers or cause various kinds of trouble.
>
> I: How could you prevent this from happening?
>
> R: We had to know their background clearly. But then we still had to watch their actions closely after being recruited. We may also be able to rectify [or redirect] them. Although they used to be bad people, they might become good after joining us. (BC-KR1)

Although recruitment patterns were different from what the top command had envisioned and at times rather loosely implemented, they were nevertheless highly selective in comparison to KPNLAF and ANS recruitment policies. Thereby the Khmer Rouge recruited a rather uniform entourage within its mid-range and rank-and-file soldiers.

This made the ranks highly permeable While rank promotion from the level of the rank and files situated below intellectuals and old military elite commanders was highly unlikely, it was the rule for the Khmer Rouge – a rule that was supplemented by a supporting factor: contacts to the top could expedite the promotion process. In addition to following certain patrimonial patterns, promotion followed screenings for especially vigorous and identified combatants and children in particular. These could rise to the level of a brigade or regimental commander. For those below the top command, a comparatively high degree of mobility seemed to be at work: "Everyone was a rank-and-file first. Later when they became more experienced, we let them become a group commander, or battalion commander. When they became an expert in fighting techniques, we would promote them to be a regiment commander" ((Laughs)) (RC-KR1). 'Work' performance counted most and was closely connected to morality. Those who proved eager to fight and fulfill all quotas and demands without complaining were viewed as good revolutionaries. For promotion, "they looked at our work performance. We were good at fighting and had good morality" (RC-KR5). It was the respective commander's duty to choose someone for promotion: "We let commanders choose as they knew their soldiers' personality clearly. They reported to us and we then promoted them to a certain position" (RC-KR1). Those who came from other factions defecting to the Khmer Rouge did not have to do 'a lot of writing' (biographical questionnaires), but many had to change their personality and be rectified to fit the ranks, especially since they had attended schools in older regimes or in the current government. This meant there was an increased need for rectification:

> Not much [had to be done after defection], only their personality mattered. They did not have to do a lot of writing or something like that. They just had to change their personality. […] Some defected to us but mostly they were literate because that side [the government] also had their own schools. (RC-KR1)

Only complete devotion to their tasks made them eligible for a promotion rank, a stance that did not allow for any sort of reluctance in combat, on duty, or in rectifying and criticizing others. Those who were not 'active' were regarded as those weak and fearful:

> We assessed how active or brave they were. Those who were brave and were not afraid of the battle were never reluctant to do their job. But there were also some inactive soldiers because they were afraid. In this case, we would put them at the back. After they have served for one or two years and have accomplished some achievement, we would promote them. (RC-KR5)

Thus, a couple of years of experience were necessary for promotion. The cadre had to prove his revolutionary stance by being 'active' by demonstrating his good morality and proper principles: "They had to have at least two to three years of experience first. When we knew that they had good morality and proper principles, we would promote them" (BC-KR1).

Transmission II: Training and Combat Preparations
The most obvious adjustment in transmission was that the blank-page operators skipped all of the formulaic trainings and opted for 'learning by doing' instead.

Although they, as mid-range commanders, might have experienced combat simulations themselves, training directives were applied in a rather pragmatic manner. Few of the training exercises, however, provided real combat preparation other than the routine of political indoctrination circles. But that training was more to re-invoke hatred against the Vietnamese:

> They studied fighting techniques, trainings of the technical part of how to use a gun, and then they received political trainings. For the political training, when they joined the battle, they would win as they were mentally full of pain because Youn killed their parents. (RC-KR1)

Most of the operators, however, did not waste any time forming their subordinates according to a Socialist norm but, instead, let combat experience do the work: "There wasn't any training, neither was there any military school. Once they reached the right age, they would go to the battle and learn in the battle directly' (RC-KR5). Whether a recruit was trained in combat simulations or directly in battle seemed to depend on the unit's organizational proximity to the center. Most of the child soldiers never received any sort of training.

> I: Were newly-recruited soldiers skillful immediately after recruitment or did they still have to go through any military training schools first?
>
> R: No, sometimes they came from a civilian background, but didn't go through any training school yet. They learned practically in the battle. Sometimes they had to go fight just a few days after their recruitment. After a while, they became skillful. This was normal for guerrilla forces. Some could survive and some died. (BC-KR1)

Those who survived proved that they were actively implementing the orders from above.

Transmission III: Rectification and Control
The operators looked to role models for instruction as well. Soldiers were rectified according to the norm – a norm of activeness in battle and fulfilling tasks. There were two types of people: the active and the lazy:

> We didn't put any blame on the [lazy ones]. They also went to battle with the others, but they were lazy. We could see this through their actions. So we could identify if they were active or not. We gave someone a leadership position only if he had a good morality, if he was active during battles, and serious with his work, and always followed commands. (RC-KR5)

But modeling was oftentimes physical rather than mental in nature. The lazy soldiers were rectified with a stick. Each candidate for a commander's rank had to go through physical education before becoming a model who could instruct others: "We educated them and let them stay with us. Unless they had been beaten with a few sticks to form them, they couldn't become a commander. Afterwards, we could ask our most genuine student to instruct other soldiers: (RC-KR4). Since correct knowledge came from above in democratic centralism, the subordinates were always reminded that if they did something wrong, they would face serious troubles or might even be killed due to their own failures in battle:

> We adopted democracy in our leadership. For example, if a Khmer soldier made a mistake, we talked nicely to them, without killing anybody. We told them that if they do not follow us, they might get in danger. This made them constantly follow us, so they never fell into any danger. (RC-KR1)

Soldiers were punished for failing to implement commands. Although there might be various reasons for this, the principle of democratic centralism maintained that failing to implement a by-definition *correct plan* meant that the subordinate rejected an order and thereby acted without the commander's consent: "If a soldier failed to bomb a tank correctly, he would be punished upon his return. His failure meant that he bombed the tank in a way that went without consent from the commander: (RC-KR2). Those who acted without the commander's consent would not be helped if they were injured and would be left to die: "We told them that if they did not follow our orders, there was nothing we could do if they got injured. But if they followed our orders, accidents would only happen once in a while" (RC-KR1). Since deviation from the collective's plan by acting 'freely' meant to put one's self in danger, the soldiers had to be instructed regularly and reminded about the possibility of being killed any time: "We didn't them instruct them only once. We did many times. We were quite strict with soldiers' discipline. If they walked freely, they might die" (BC-KR1). Political indoctrination involved talking about the morality of a soldier, which encompassed just about everything. It started with rules such as not to steal property and to have the correct weapon and appropriate bullets with you at all times:

> We talked about soldiers' morality. While liberating a village or a location, we mustn't steal people's properties. We must only think about protecting the people. We must take the right bullets with us. For example, if our gun was an AK, we must take AK bullets. If it was a SKS, we must take SKS bullets. We mustn't think about stealing people chickens or ducks. (RC-KR2)

Everything was considered within the realm of proper morality. If a soldier made a mistake in handling his weapon and ammunition supply, this constituted a moral misdeed, as did hurting civilians. Rules were strict and applied to each act that may signify lack of Socialist favor in its execution. Any failure that would disrupt the collective movement had to be prevented:

> One more thing we had to bear in mind was that we had to make sure we finished our breakfast and went to toilet by four or five o'clock at the latest. We were not allowed to stop to pee while walking, or our walking line would be disrupted. (RC-KR2)

Enjoying yourself within the collective proved highly dangerous: "Back then, no one dared to drink, sing, or dance like now. If they did that they would be killed" (BC-KR1).

Transmission IV: Surveillance and Control
Khmer Rouge control permeated right into actual battle. While some other commanders in the field did not have a close grip on events during combat, the Khmer Rouge apparatus had an eye on everyone in detail: "They could not hide in front of us because we would eventually know if they had done anything wrong" (BC-KR1). The Khmer Rouge opted for on-the-spot surveillance that was facilitated by each rank checking up on its direct subordinates and vice versa. For example, those who tried to evade being involved in battle were simply shot on their way back (RF-M1), since a section commander always attended the battle (RF-M2). Combat surveillance was tight-knit, and so was post-combat reworking and processing. Criticism sessions strengthened the close-knit control structure by adding peer pressure as described in the section on the anti-intellectual intellectuals. Any bad habit or trace of potential indiscipline in a soldier's social background was supposed to be constantly reflected upon in order to purify the subject according to a revolutionary morality. As a result, livelihood meetings followed each combat operation: "After fighting in a combat, we would identify our strengths and weaknesses" (RC-KR5).

Only one commander mentioned the chance to circumvent command by playing with the word for Wednesday (*thngai put*), as the Khmer word *put* also means 'to pretend':

> We pretended to be sick or have a stomachache. In the military, we could only relax on Wednesday. Civil servants can relax on Saturday and Sunday, but soldiers can only relax on Wednesday, pretending to be busy because children were sick and so on. This was called an excuse. Soldiers always have to fight, regardless of time, but civil servants can completely relax on Saturday and Sunday. But for soldiers and police, they have to do their job when the enemy arrives, even at night. I used to ask people when soldiers could relax, and they said on the weekends. I said they could relax only on Wednesday. They said Wednesday is a working day. I said soldiers could pretend to be sick or busy sending children to hospital. (RC-KR2)

Only a few respondents mentioned the possibility to refuse orders. The Khmer Rouge apparatus did not leave much, if any, space for disobedience, since any deviation in conduct was met with drastic means for rectification and, eventually, execution.

Chapter 6
The Rank-and-File Soldiers

Rank-and-file soldiers were shrouded in myths. The upper leadership of the resistance spun many theories about their nature and regarding what their mindset might have been. 'People', as an intellectual commander contends, simply love to engage in war, and recruiting new soldiers is seen as a necessary part of that:

> The psychology was really in wartime (1) you (.) you see people do things that they normally don't do (1) rape (.) stealing, you know (1) killing (2) and it (.) when the people (.) I don't say that the people love to do war (.) but for them (.) because it is wartime (.) it is normal that they are recruited (.) to be a soldier of the resistance (.) to be a soldier of the resistance movement (1) so (.) there is no problem at all, you know. (RC-KP1)

Although for most respondents from the rank-and-file level, it was indeed 'normal' to become a low-ranking soldier and not a commander, being recruited for the guerrilla resistance was not a 'natural choice' for them. While it was logical for individuals from a low social milieu to enlist as foot soldiers serving at the front line, they neither loved to engage in war nor exhibited a twisted psychological understanding of war due to the conditions they faced during 'wartime'. Rank-and-file soldiers were not simply characterized by a complete breakdown of ethical conduct, making them kill, steal, and rape at random. Reasons that young men applied to become a soldier (or a 'messenger' in Khmer Rouge terminology) in the resistance were manifold and will be discussed in the section on access to the field. Moreover, dealing with war, fear, battle, and discipline from above will be the topic of the sections on discourse and power practices.

Habitus Formation and Lifecourse: Non-Communists

All rank-and-file soldiers within the two non-Communist factions were from rural areas and came from peasant families. Some had been construction workers or even low-ranking soldiers under Lon Nol. They received little to no education and oftentimes belonged to the class of 'base people' under the Khmer Rouge. When talking about their families, they usually simply stressed again and again that they were "normal": peasants making their living with subsistence farming who were too poor to send them to school. Their age at recruitment was a decisive factor for their habitus. Most of the very young recruits adapted to an 'extremist pragmatism' of war times; those who were older framed war and the life of a soldier in terms of a vivid animism and spiritual classifications. Thus, within the rank-and-file soldiers, the main difference between habitus groups lay in the age of the soldiers and whether they had experienced something other than a society at war.

During the Khmer Rouge Regime

Most of the rank-and-file soldiers experienced an increase in their social and political status during the Khmer Rouge regime. As 'base people' they were regarded as closer to a pure revolutionary mentality than the deluded 'new people' from the cities or 'exploiting classes'. This, however, did not mean that they were actual perpetrators during the regime, as many 'new people' suspected them to be. Most experienced a merely symbolic increase in social value – a change in symbolic status, which, due to resentments against townspeople, many people enjoyed at the very beginning. But in the end, many 'base people' also suffered under the new regime, with all its food shortages, lack of medication, bans on religion, co-operative working and dining, and constant vigilance due to Angkar's system of terror. After a year in power, more and more 'base people' were imprisoned in the security and re-education centers for 'rebuilding' their mind or even for execution. While their chances of obtaining a position in the party were much higher than for 'new people', which meant that they could make use of some comforts and oftentimes evade execution, especially during the early months, the security apparatus increasingly turned an eye on 'unrevolutionary' sentiments among the revolutionary base as well.

The younger children remember only food shortages and separation from their families, and oftentimes they did not have anything to share about how their parents raised them. Therefore, the first question about their family was difficult to answer and inquiries regarding how their parents raised them did not make much sense to them:

> My parents did not raise me up much. During that time, they could not do so because it was a period where everything was organized in cooperatives. We had to eat collectively. After '79, my dad died, so he could not raise me up as well. When I turned fifteen or sixteen, I was forced to join military. (RF-KP2)

Many did not experience family life, and, even after the regime's demise, they could not return to live with their families because many of their parents and siblings had been killed.

Access to and Mobility in the Field

Only very few joined the resistance for the purpose of fighting the Vietnamese. The official master cleavage was increasingly cited as the reason that higher respondents joined the resistance, but it became almost obsolete among the lowest rank-and-file soldiers. When asked about why they joined the resistance, most of them told stories about seeking for food, shelter, resettlement, or simply about bumping into a camp by accident. Reference to fighting the Vietnamese was often just ironically given or was reduced to hearsay about the Vietnamese being everywhere. Most of the rank-and-file soldiers came from the northeastern part of Cambodia, where famines were the worst after 1979, fights with the Vietnamese military were fierce, and aid distributions in close proximity: "We heard that we would have nothing to eat if we stayed in the interior. And we also heard that there was a humanitarian organization over there, so we decided to go" (BA-KP3).

For the most part, it was the better food situation and not the resistance that drove people there. The proximity to aid distributions pushed people to the border as well. Some arrivals immediately opted for resettlement, but most headed back to their villages if possible at the beginning. Resettlement, however, was mostly an option for higher social milieus and not the peasantry, due to a mixture of not being suitable for resettlement and not being willing, since they only came for food. Rank-and-file soldiers, by contrast, oftentimes did not even know what that "third country" was and thought it could be Myanmar, Thailand, or Singapore. The number of refugees coming in dropped year by year because the situation stabilized in the interior, but it peaked again when the Vietnamese launched their dry-season offensive and conscription for K5 in 1985. Some fled conscription by the state military and then ended up in a camp as a soldier for the resistance (e.g., RF-F1).

Others went to the border to make a living by smuggling goods to the interior. When conducting cross-border business, Thai soldiers and policemen constantly harassed them and took bribes for the passage of goods. However, if they were wearing a military uniform and protected by their own official status as a member of the KPNLAF or ANS, they could easily conduct business without harassment and paying bribes (SC-KP1). Having a weapon protected them from Thai intervention into their business, and having a commander with similar interests meant that they could institutionalize smuggling with 'military backing'. In comparison to other smugglers, soldiers had a lot of money and fancy clothes, as one soldier pointed out when explaining his reasons to join:

> During that time I was twenty-three to twenty-four years old [indistinct] and moved from the province to the city leaving the city, in order to become a smuggler along the border. When we arrived there, we saw soldiers having a lot of money and being impressively dressed and ordinary people, who were no soldiers, looking powerless. That means if we became soldiers, smuggling as we wished would be easy. (SC-F1)

Not everyone went to the border intentionally: Some even ended up there due to erratic flows of people after the end of the reign of the Khmer Rouge, when everybody was looking for their relatives and others walked home by oxcart or by foot after relocation. Still others only went to the camp to get some food and planned to return home immediately afterwards, but then they got stuck there: "My friends were just like me. They got stuck there [at the camp], and then became soldiers" (GC-KP1). Some people report how they were threatened to stay by the resistance: "They had the saying that when we go to the interior, there were only mines and they would hammer our nuts" (SC-F1). Many of the lower ranks were tricked into thinking that they would be suspected of being a former Khmer Rouge when they returned or that they would be killed as resistance soldiers upon their return. Some knew rank-and-file soldiers from before, but this was only very rarely the case, and if it was, it meant that they had someone to drink with rather than a resource for their own promotion: "We went via network, so we all had fun" (SC-KP1). Many report that they first tried to make a living in the camp. In addition to the distribution of food, people could do some cultivation on the land. However, in the end, cultivation was too costly, since they had to give large parts of the crops to the section leader of the camp and plots were small. Other sources of income included gathering food outside of the encampment (taxes at pass-ways, being hit when caught slipping through fences), employment in camp administration or the international community (mostly accessible for loyal followers of the resistance only), or work as a barber, blacksmith, seller at markets, or as a smuggler (difficult without being a soldier) (cf. Reynell, 1989, pp. 92–123).

In the end, however, job opportunities for men were scarce, and many were pushed by their families to earn some extra food by joining the military. This was the by far easiest way to get more rations. Due to the UN distribution system, which gave no rations to men above a certain height, many men married a woman to get her rations, or they joined the military. Many state that the distribution system pushed them into resistance if they were not married:

> There were no any other choices besides becoming a soldier because rice was given only to ladies. Guys like us were not given anything, so we had no choice but to be a soldier so that we could have rice to eat. If we didn't become a soldier, we would have to get married in order to get rice. If we were single, we had to be a soldier. (SC-F1)

Some younger male refugees tried to pretend to be a woman when getting in line for the rice distribution. But that trick ran the risk of being revealed and punished severely (GC-KP1). Rations received from commanders were not always sufficient to sustain life, either. Only those who managed to become a group or section leader were better off, and those who managed to slip into a commander's entourage also lived a comfortable life. Others still had to earn something extra.

'Spiritualists' and 'Pragmatists'

The main divide within the lower ranks lies between recruits who were young when they became soldiers for the first time and those who were adults at least twenty or more years of age. While the habitus group of 'spiritualists' was rather old, the 'pragmatists' were young adults (fourteen to eighteen years) or even children (eight to fourteen years) when they were first involved in military socialization. While their social positioning and background were largely the same (lower class, poor peasantry), their lifecourse and participation in the field was very different. From a synchronic or structural view, their social position and early 'milieu' within their family of origin are the same, but from a lifecourse perspective they differ a good deal. The 'only' material difference in habitus formation between the two groups is their age at recruitment or whether they faced displacement and dangers of war instead of village life. This divide, however, is not necessarily merely a generational divide. It is possible that someone was forced to join an armed group during the Lon Nol regime at an early age and had a lot in common in his habitus formation with someone who was forced to join the CGDK resistance twenty years later. Both experienced large parts of their socialization within an armed group. Although the 'pragmatists' typically joined as children under the resistance, there were some child recruits from Lon Nol as well. 'Spiritualists' typically experienced a childhood with a family before the Khmer Rouge took power, or in some cases even continued to live with or have contact to their families even after the Khmer Rouge. Both of the groups have a few disciplinary practices in common. Before discussing each group's distinctive discourse and practices, the concept of disciplinary self-techniques and some common characteristics in disciplinary practice shared among these groups shall be outlined.

Self-Techniques and Strategies of Flight

Rank-and-file soldiers disciplined themselves. Their arbitrary and socially constructed position in the field as the ones who were in the position to do the actual fighting was either reinforced symbolically using self-rectifying techniques or power was undermined using strategies of flight. While no rank-and-file soldier would challenge the fact that he was 'in the position' to be a combatant, they all had to deal with their own fears and ethical reluctance. Depending on their own habitus formation and symbolical universe, these men developed techniques for overcoming fears and reluctance (self-rectification) and criteria for obedience. Their ideas of being a soldier and regarding leadership determined the conditions of resistance and flight within their unit. The classificatory discourse in this case structured rules according to which under unbearable conditions either the self was adjusted to the conditions or the conditions were changed to match the demands of the individuals. Some disciplinary techniques were shared among all rank-and-file soldiers and will thus be discussed first. Others techniques, however, depended on their habitus formation. Yet all of the techniques shared a basic mechanism of symbolic violence, which means that an unequal position became legitimate and the subjugated adapted to fulfill his symbolic position with all its 'necessities' (Bourdieu & Wacquant, 1992, p. 143).

Fear

> Shooting during battle is different to what you see in the movies. We cannot always shoot well and technically properly. Some soldiers were so scared that they did not even know where the trigger was. (GC-KP2)

Disciplinary practices revolve around obedience to commands during battle and beyond. In addition to making soldiers obey orders that they normally may not follow, shooting in battle and fear of all kinds of dangers and death were at the center of such disciplinary techniques. Soldiers in combat are far from loyal automatons, no one interviewed had been free from overriding fears when confronted with actual combat situations. Strongmen staying close with their own soldiers in battle saw their fears: "Some peed their pants without noticing" (BC-F3). All theory and all preparation melt away when you are confronted with real enemies trying to kill you or whom you have to kill.

For some, only direct surveillance and threats during combat helped to make them fulfill their 'duty':

> Honestly speaking, (2) when we actually had to shoot at our enemies, it was very different from what we learned during the training. Once the enemy fired, I lowered myself to the ground and only went forward when my boss kicked my ass. Only then I was following what others were doing. (GC-KP2)

There were few chances to evade combat situations when there was a commander monitoring you, threatening to shoot you if you tried to avoid real shooting. Even if they managed to evade their group, running away was not a real option either:

> For me, when I was asked to shoot, I got scared. But we had nowhere else to go. If we ran away, we would get lost in the woods and die anyway. We had to stay with the group. [...] During that time, even a hero got scared. (GC-KP2)

Before discussing techniques of different habitus groups within the rank and file to circumvent orders or deal with fear, one technique shall be discussed that was mentioned by almost all respondents across all groups: routine.

Routinization

The simplest technique for dealing with fears was exposing the individuals to their own fears and actual combat again and again in order to develop a routine. To do so, many soldiers went to battle and shot "randomly" to get used to it (GC-KP2), pulling the trigger without looking and without intending to hit any target introduced them, but for sheer practice in battle:

> I: How did you deal with your fear?
>
> R: Shooting! Shooting back and forth! Although I did not even know whether I would hit any target. (CC-KP1)

Others likened it to parachuting: "It's like with parachute jumpers. They are usually scared before jumping the first time, but after the first jump, they would not be scared anymore. After five to six times practicing, jumping with a parachute would be something normal for them" (BC-F3). Reluctance to fight was overpowered by routinization, a very effective tool for dealing with extreme fear. In the long run, as most stressed, fighting became a habit, in which the pure act of shooting cooled them down and, after a while, made them brave:

> R: To avoid fear, I shot. After a while of shooting, we were no longer afraid. Without randomly shooting for a while, we couldn't calm ourselves down.
>
> I: Were you still afraid later on?

R: Not anymore. I don't know why. At first I was very afraid. But after shooting for a while, I wasn't anymore. I became so brave. (BA-KP3)

While stressing that routinization was very effective, it was supported by various means during combat and beyond, which varied across the ranks and groups. After volleying some bullets, they adapted to their own 'duty' and to the duress of a combatant.

Ethical Guidance of Bullets: Meeting Friends, Neighbors, and Brothers in Fear
Combatants made a difference that mattered because it set them above "brutish others" (GC-KP2). By introducing an ethical code they were to follow in combat, they legitimated their own actions as being ethically pure(r) in comparison to others. Many, for example, said that they were reluctant to shoot at "friends" or simply people they knew from their hometowns and villages:

> During the time we were fighting with each other, it happened that we knew each other. We had friends like you and me, near to each other, knowing each other, but now fighting at the same area for different factions. Sometimes we would simply recognize our friend's voice, so we asked each other to stop shooting and relax. Although being told to shoot at each other, we did not hurt each other seriously. The exchange of fire did not happen excessively although we shot bullets of all types. However, sometimes we took fighting serious and sometimes it wasn't taken that serious. (CC-F1)

If the whole battalion or fighting unit was reluctant to fight, there were various ways to prevent actual fighting. The easiest way was to pretend that they did not meet the enemy in the first place:

> For example, when we met people we knew on enemy side, whom we did not want to shoot at, we would simply pretend that we haven't bumped into each other. How could we fight if we never actually met each other? ((Laughs)) (BA-KP3)

But if they were ordered to fight a certain group at a prescribed area, the upper echelon was often listening via radio transmitters. However, they could only listen. Therefore, units could negotiate with the enemy and stage fighting by randomly shooting bullets into the air that could be heard by the command listening at the transmitter and by villagers nearby (BA-KP3). Others showed mercy towards those who were brothers in fear and hid themselves somewhere during battle. When they bumped into them, they felt pity towards them and spared them from their own fate: "I never shot at those trying to hide themselves out of fear" (GC-KP2). Every respondent had some sort of ethical guide at hand, and most said that they did not kill Khmer. While killing Vietnamese or 'Youn' was praised by most, some even bragging about killing a couple of them barehanded (e.g., RC-F2), they tried to make clear that they avoided shooting Khmer. However, self-defense was often raised as the main reason why this was not possible in actual combat in the end. Or, the soldiers depersonalized the actions with a phrase like "bullets have no eyes", especially when civilians were shot (cf. ICRC, 1999, 26). While these ethical guidelines might in some cases have actually affected action in combat, their main function was in a symbolic increase in the value of the soldier (especially in front of the researcher), oftentimes painting a picture of how brutish others were:

> Some commanders were so brutal. When an enemy got shot in his leg and might even survive with some help by others, they would shoot him right away. For me, by contrast, I would ask the wounded whether they are Khmer or Vietnamese. If he was Khmer, I would help him to survive. If he wasn't, I wouldn't help. That was my approach. (GC-KP2)

These ethical "approaches" diverted attention from the act of killing someone towards an act of mercy saving some of those who were otherwise condemned to death. Employing the ethics of mercy towards certain groups was a technique of both self-rectification (soldiers lowered their own reluctance to kill) and avoidance at the same time (they avoid killing certain people they were commanded to kill). One habitus group within the rank-and-file soldiers, however, saw no place for ethics in war: pragmatic clients. Although they themselves talked about fear and mercy in front of some of their foes, they stressed that being a soldier or military commander meant that they were to act in a space that was not ruled by any kind of ethics other than that of 'extremism'.

Pragmatic Clients

Members of this habitus group oftentimes looked back on decades of membership in more or less small-scale armed groupings at the lowest ranks and/or as smugglers along the border. Throughout most of their lives, they lived with harsh insecurity of basic survival, dangers of all sorts, and daily sufferings. In order to survive during the war, they looked for patrons to take care of them by securing their businesses and giving them protection in front of the Thai border police and military. Therefore, in order to be able to go on with business, many joined the military: "As doing business with the Thai was tough, I joined the military" (SC-KP1). Most of them, however, were subordinates of strongmen or took shelter within their units, since strongmen engaged in large-scale organized smuggling to finance their followership and could provide well as a patron. They were the most potent caretakers at the border and most likely could provide well for their subordinates. The group's discourse and practice was characterized by a 'pragmatism' or even 'extremism' that sought for nothing more and nothing less than becoming a client under a patron providing for basic needs. 'Pragmatic clientelism' structured their conditions for loyalty, their techniques of rectification, and strategies of avoidance.

Classificatory Discourse I: Conditions for Loyalty to Patrons

Most clients stayed with their commanders simply to secure at least some food to survive. However, they might have joined for some other reasons, as a child soldier highlighted who simply liked the soldiers' fancy clothes and impressive guns:

> As I have said, before I went into the battle, I saw soldiers carrying guns, and they looked so fancy. But when I experienced battle by myself, I wanted to resign afterwards. However, I was not able to. Although they had allowed me to quit, I wouldn't know where to go to get food to eat. Hence, I had to tearfully stay. Since then, I became starved just like during the Pol Pot regime. (SC-F1)

Without a military patron, many feared that they would not be able to secure sufficient food to survive. Finding someone strong who could take care of their basic livelihood was the main strategy for the 'pragmatic clients'. A commander was not expected to provide much more than food. In a rather ironic tone, one soldier emphasized that all he was hoping for was food and to shoot some Vietnamese: "As long as we had enough rice to eat and Youn to shoot, we would not go anywhere else" (SC-KP1). All they were looking for was securing their very own survival, though they did not feel very good about getting everything from patrons in the camp, patrons in the international aid community, or anyone else. Many felt like no one actually cared about their lives: "I only thought of my own basic survival. We were wearing their clothes, using their guns, and eating their rice. If we died, they wouldn't give a shit on it" (BG-F1).

Since some felt that they were worth nothing in the eyes of their patrons, it basically did not matter for which faction they fought. They joined purely for pragmatic reasons; neither politics nor ethics mattered. Therefore, the main difference between all of the factions (including the government forces) oftentimes lay in the amount of food rations that were provided: "The difference between all three factions I was in [Khmer Rouge, the incumbent KPRAF, and ANS] lies only in sufficiency of food" (RF-M1). Expectations were low and some thought they would die soon anyway: "I was ready to die because I thought I would die soon anyway – either by sickness or something else. At least, there could be someone to organize my funeral" (RF-KP1). All they needed was "enough food to eat, proper clothes to wear and sufficient financial support. These would make soldiers happy" (RF-KP2).

Practice I: Defection

Pragmatic clients never mentioned that they joined for any political reasons (e.g., to support any political group or expel the Vietnamese), but only mentioned 'incentives' such as food, security, and shelter. Thereby they fit into Jeremy Weinstein's description of 'opportunistic recruits' seeking for material benefits only, who are also unlikely to remain in a faction when these incentives cease (2007). Defection became a common strategy among the 'opportunists' within the CGDK. However, these 'opportunists' did not seek to maximize their gains by changing their patrons; instead, they looked for someone who was likely to be able to guarantee sufficient means for them to survive (subsistence). Therefore, an ongoing lack of any of these goods (food, security, shelter) meant that they looked for a strategy to flee. Defections to another subgroup within the respective faction, another faction of the resistance, or the government were the most common. Many respondents from these habitus groups had already been members of various factions throughout their career as rank-and-file soldiers (e.g., RF-M1; RF-M2), and the rest defected at least once. While members from the habitus group 'spiritualists' did not defect at all, and many even believed it was impossible to do so, the pragmatists, in contrast, stressed how easy it was.

Some defected because their commander was brutal and thereby threatened their physical safety: "If a commander was too brutal to me, I could not stay with him" (SC-KP1). Others defected because they made a mistake that would have been punished if they had not fled: "Were you afraid of being punished for misconduct? R: No, I wasn't because we could run away anytime and simply went to a different place if we were to be punished" (RF-M2). Sometimes the former commander would chase them:

> Sometimes, we simply ran away. Once caught by the other faction, they would ask us where we came from. Then we would simply say that we came from over there and that we wanted this new faction because we didn't like the politics of the old one. We could just say anything in order to survive this. However, running away could be risky as well. If we ran away because we made a mistake, we would be chased. (RF-KP2)

However, the likelihood of being caught was rather slim, and paying small lip service to the group's political goals was typically sufficient to be accepted within their lowest ranks. Some never heard of anyone being caught while trying to defect. Changing their names made it rather easy, and combat operations to the interior presented a good opportunity to flee from their group: "I: What happened if someone was caught who tried to defect? R: I don't know because they never caught anyone. When we defected, they would not even know. We always changed our names" (RF-M1). However, defection only ameliorated some of the worst effects of a life at the lowest level of command. Although there was some "competition" among commanders (RF-M2), changing the patron did not facilitate any sort of vertical mobility that would lead to a fundamental change and improvement in their living conditions. Resistance to power by defection was limited to lowering its most threatening violations of the clients' basic survival.

Classificatory Discourse II: Ethics of Unethical Extremism

For the pragmatists, unethical extremism lay in the nature of being a soldier or military commander. To be a soldier meant to be extreme and bad. One respondent tried to explain the divide between soldiers and ethics and good conduct with reference to becoming a monk:

> Soldiers were not good unless they became monks so that they don't kill anyone anymore. The same applies for commanders in any kind of regime. You cannot be a commander without being extreme. If you want to be good, become a monk.

A commander had to be extreme; otherwise he cannot be a good commander. Death and life must be closely together. (SC-KP1)

Soldiers are always about to kill or to be killed by someone else. Pragmatists lived in a constant state of emergency, and since their early childhood that had become a guiding principle within their judgments and practices. Moreover, their respective commander had to be bad as well. It was assumed to be in the nature of being a strong commander to be ruthless. Commanders were *expected* to use their power arbitrarily, as Lindsey French highlighted when discussing sentiments among rank-and-file soldiers within the CGDK: "The arbitrary exercise of power was not considered, in itself, bad. It was simply the way power was. People just wanted to be on the side of the most powerful, because then they would be protected" (French, 1994, p. 177). However, this sentiment or discourse was only spread among the pragmatic clients, who nevertheless would defect if a certain threshold of brutish conduct was reached. Only for them, misuse of power that ignored all kinds of rules and ethics was accepted to a certain degree, and seen as the extremist reverse side of the nature of power. Hence, only for the pragmatists, commanders had to be bad and indiscriminately hurt people: "Commanders were rarely good. Commanders could not be good. They simply could not be that way. They had to shoot, chop, and stab all the time" (RF-KP2). Kicking and using their subordinates was considered normal because: "We were their pedals" (SC-KP1).

Buddhism or religion as such was wasted when provided for a soldier without any moral awareness: "I: Did you receive morality trainings? R: NO, because soldiers did not have any moral awareness. Buddhism allowed us to do some chants, but soldiers only knew about killing" (SC-KP1). The same section commander went on to make a joke using a double meaning of 'feeding someone' (feeding monks – feeding enemies with bullets): "Monks also gave me some *Yorn* [for protection] in battles and in villages. Because we also gave some food to the monks. But there was no monk in the forest ((laughs)), so we could only give some food to the Youn" (SC-KP1).

In great contrast to many others within the rank and file, the pragmatists saw no use in magic and any kind of 'superstition', as they would call it. When asked which kinds of magical protection he used during battle, one rank-and-file soldier stressed that for him protection was a matter of a correct use of hills and trees: "These [religious] instruments were just superstitions. For me, I totally relied on small hills and trees for protection. That superstition was just some people's belief. However, it did help them to deal with fear" (RF-KP1). A guard of a brigade commander put it a bit more bluntly, making the belief in magic responsible for dangerously careless behavior during battle: "Some of them were really stubborn. They were told not to walk past a certain field as mines were buried there, yet they did not listen. These people ended up stepping on a mine and dying" (BG-F1). Oftentimes, the pragmatists told similar stories of people walking across battlefields believing that they were bulletproof. The same guard ridiculed:

[When arriving in Laos after fleeing from the interior], I was arrested and transported to a police office. Then they gave me a bowl of magical water and told me drink it. They said that if I came there with good intentions, I would survive drinking it. But if not, I would die. Then I said I would like to drink it, and unsurprisingly I survived after drinking it all. (BG-F1)

Practice II: Becoming Extreme

Pragmatists did not use spiritual and magical instruments in order to adapt to dangers and fears they faced as soldiers in battle. None of the respondents had any tattoos on their bodies (in contrast to many others), and they all emphasized that magical instruments made little sense to them. Even if they had some instruments, they did not feel very good about wearing them: "I had nothing [magical for protection] as I did not use any religious instruments. I sometimes even threw them away because I did not feel it was convenient to wear some" (RF-M1). Many only mentioned routinization as a way of dealing with their fears. At the very beginning, only fleeing from battle situations could help: "At first, I ran away because of fear" (SC-KP1). But then routinization set in. and shooting some bullets made them less afraid from time to time: "After a short while when we exchanged some fire, I was not afraid anymore. We were most afraid before meeting the enemy, but not later on" (SC-KP1).

Widespread among the pragmatist was one disciplinary technique that made them fight any kind of enemy they would face: drugs and other substances, primarily alcohol. Many of the respondents turned up at the interview more or less drunk, and they also reported about using alcohol to make themselves fight without fear. Or, as a section commander reported: "We drank wine until we were able to see [approach and fight] all kinds of giants. That made us extreme. Without wine, we could not do it". Therefore, he had to constantly carry a bottle anywhere he went: "I always carried rice wine with me. No matter what happened, I had some with me" (SC-KP1). For most of the soldiers, their commanders were far away and did not know what they actually did during operations. Therefore, they stressed, it was possible to do anything they liked: "To make it short, higher officials didn't know anything about what happened during combat operations" (BG-F1).

Ethics of mercy could be found among the pragmatists as well – in spite of their habitus of being 'extreme' and fearless. However, underneath a certain degree of swaggering about their reckless attitude, most mentioned a similar system of mercy during combat. While a section commander first bragged about having fun killing Vietnamese: "Being a soldier was fun because we could shoot Youn" (SC-KP1), the same soldier later on mentioned how he felt pity for the enemy, who in his eyes also just followed some orders forcing them to fight like he and his comrades had. In this regard, he considers the Khmer and Vietnamese to be quite similar:

We really did not want to shoot Khmer soldiers since we were the same. But if they shot first, we had to defend ourselves. We really did not want to shoot. We even did not want to shoot the Vietnamese soldiers. [...] We were all forced to shoot by our commanders, and so were the Vietnamese. But if we refused an order, both of us would get into troubles. (SC-KP1)

This is a typical sequence: While soldiers recited a master cleavage and widespread sentiments against the enemy, they felt connected in fear and acknowledged being in a similar situation to that of their own enemies. Both sides were just people who had to follow orders and fear death. Not shooting at those hiding in fear was a way to deal with this emotional paradox without 'getting into trouble'.

Ethical Spiritualists

Although suggested by some (e.g., E-US1; French, 1994), having sufficient food, shelter, and security was not the only binding norm for soldiers. Another group, which could be termed 'ethical spiritualists', structured its disciplinary system differently. Jeremy Weinstein (2007) might call members of this group 'activists' who do not seek the private benefits of joining the resistance but join a resistance for ideological reasons and invest in the group. This divide is highly simplified, and already the pragmatists proved that they joined neither as a means of maximizing their benefits nor for any other solely opportunistic reason. The main difference could be found in the lifecourses of the two groups, with pragmatists' upbringing and socialization occurring more in the realm of armed groups rather than in a family setting. Still the pragmatists classified themselves as 'unethical extremists', and their behavior was thereby guided by the 'ethics of the unethical' and a norm of basic subsistence. The 'spiritualists', in contrast, applied a whole set of spiritual norms and ethical demands to structure their rules for compliance towards their own commander. These norms centered around 'respect': As long as their commander respected them, they were more willing to accept and bear material hardships for him. If not, they did not defect but, rather, believed that fate would serve as a judge for the bad commander or took action themselves by taking revenge for humiliations.

Classification I: Spiritual Potency and Fate

The spiritualists shared with the strongmen and some mid-range operators the belief that the world was animated by spiritual powers, with rules that needed to be followed in order to gain strength. They carried a strong and often-cited 'Buddhist' belief: "It wasn't necessary to teach us on Buddhist ethics. Everyone was aware of Buddhist teachings on their own" (SC-F1). In their view, such beliefs served as a common guide for every Khmer in the country, and even more for soldiers who could make use of religious powers and instruments for their own protection: "Buddhism is common in our country. A hundred per cent of our soldiers believed in it. In Buddhism we have bulletproof instruments and teachers. That's why soldiers all believed in Buddhism, even more than ordinary civilians" (SC-F1). But power was not just found within instruments, magical spells, and charms. All

Fig. 6.1   Magical handkerchief. Source: picture taken by author, September
              2011.

of these instruments only functioned in combination with a spiritual soul of those
using them. Only real spiritual belief made them functional and granted the soldier
supernatural powers: "Those having a spiritual soul in their body were bulletproof"
(BA-KP1). People dying in combat despite wearing magical instruments, thus,
had no real spiritual soul carrying the powers of the items, or they simply trusted
the wrong item or wrong magician. The system never failed: instruments in
combination with a spiritual soul made the soldiers 'bulletproof'. But if one
element was missing, the protective effect stopped working, and they were killed.

Practice I: Spiritual Items and Discipline

As already outlined with the strongmen, there were plenty of items that were used
for protection. Beyond tattoos there were charms to ensure happiness and safety
(*mun keatha*), Buddha necklaces and statues made from bones, solid pig spurs,
and frequently also handkerchiefs (*yorn*), as pictured below during an interview
with a monk:
    Such items were believed to radiate power and strength, or even to make
soldiers invisible during combat. Others endowed commanders with the power to
yell more loudly as a useful tool for a military leader. Each respondent from this
habitus group was tattooed all over his body.
    However, practices served as mediators between the soul and the spiritual items
and introduced the possibility for the agent to actively influence his own fate.
Thereby soldierly discipline became structured by spiritual ethics. These practices

followed certain rules taught by monks and magicians. In order to become skillful and survive battles, the soldier had to adapt his behavior to this ethical system, which was sometimes simply a matter of dietary and hygienic regulations: "A skillful soldier has to follow certain formulas and lessons" (GC-KP2). These rules were explicitly ethical, such as not to threaten or shout at someone, not to steal, not to rape someone's daughter, or not to do harm to civilians. But they could also be more related to dietary and hygienic regulations, such as not to eat pork or not to wear your amulet during a visit to the toilet or not to urinate and defecate at the same time (Maloy, 2010). Breaking these rules meant being killed in combat, and being killed meant that some rules were broken:

> Morality had its relevance in this superstition. Those who learned magical rules and conduct could not rape other people's wives or daughters, or they would die. They really believed in this. This was a way to educate people to have a real belief. Some people believed so much in this that they had tattoos all over their head and their entire body. (BC-F3)

By using religious protection or rituals to lower their fear during combat, a situation of force became legitimate, and the subjugated adapted to fulfill his symbolic position with all its 'necessities', a process Bourdieu would call 'symbolic violence' (Bourdieu & Wacquant, 1992, p. 143). In this regard, spiritual systems may function as symbolic violence, in which soldiers accept their position and possible death as their own 'fate': "No one was afraid. It was our fate" (GC-KP1). In the end, however, "life and death was already determined by fate anyway" (GC-KP2). Unfortunately, fate only revealed itself after death.

Those who were killed broke the rules that protected them from death. A rank-and-file combatant recounted two events that proved the effectiveness of magic to him:

> During a fight with Youn in Kratie, there was a guy, who was bulletproof. But later on he cursed me, so I said he would not live much longer. I only said that. One day, he betrayed the commander and pointed the gun at commander [name], so our group chased and shot him dead in spite of having tattoos all over his body. This means that only as long as he was strictly obedient to the rules, he was bulletproof. That's why I believed in that. A nun from Kulen Mountain gave me a Buddha necklace statue and told me to strictly follow the rules such as not raping other people or doing anything morally unacceptable. As long as I strictly followed that, I knew that it would help. For example, I was able to foresee upcoming obstacles and dangers, which is why I was able to be more careful. Normally, I would have walked in front during a combat operation. But since I knew that I would face some kind of danger soon, I was more careful and stayed behind. If something strange happened to me as a sign of danger, I would reject the order from the commander because I was afraid that I would be in trouble. I strongly believed in such superstitions. (RF-F1)

Hence the spiritualists did not follow orders that ran contrary to these magical predictions and precognitions.

Classification II: Respectful, Humble, and Serious Commanders

In stark contrast to the pragmatists, having sufficient food was not of primary interest to the spiritualists when judging whether a commander was good or bad. For them, lean times were part of a soldier's life and therefore nothing to worry or complain about:

> As for food, we all know that during war soldiers cannot be like ordinary civilians in the country. We didn't care much about having food. If we had, we would eat. If not, we wouldn't. When we had no food, we could also eat some nonpoisonous tree leaves. (SC-F1)

When talking about a good commander, they characterized him in terms quite the opposite to those used by the pragmatists. Extremism was not expected as a central behavioral trait, but variations of gentleness, humbleness, and seriousness were constantly highlighted (e.g., SC-F1; SC-F2; RF-F1; GC-KP2). Being humble or gentle primarily meant to show respect towards their subordinates:

> If a commander did not show us his respect, we lost trust in him and our morale went down. For example, despite all the victories and achievements we reached during many battles, some commanders still ignored us. This made us upset. In this case, I would reconsider my efforts for them. I had done a lot of things for them, yet they did not give us any good returns. They did not give us love. (GC-KP2)

Similar to the strongmen, the spiritualists' view on a good commander was that of a model in spiritual and soldierly ethics. He had to stick to spiritual rules embedded in his role as a role model centering on humbleness, seriousness, and respect shown towards his subordinates. But strongmen emphasized bravery and strength alongside spiritual codes, which mainly served to enhance their own strength and power. For them, these ethics centered on gaining strength by making use of magical powers. The spiritualists, in contrast, did not mention bravery and strength as being part of a good commander's model but connected good conduct to respect and humbleness, being spiritually codified and secured. Deviations from this model of good conduct were punished by magical powers that deprived the commander of the strength necessary for his own survival or (in the long run) by their subordinates during combat missions.

Practice II: Cosmic Revenge for Disrespect

Not many of the spiritualists were upset with their commanders due to lack of respect to the point that they defected. None of the respondents interviewed ever left their commanders. Many even stayed loyal to him until the day they were interviewed. Instead, perceived humiliations in which soldiers felt they were not being respected well enough could be answered with a practice Alex Laban Hinton worked out as a central feature of Cambodian culture as such (Hinton, 1998,

2004, 45–95). While I would downplay its relevance to certain groups, in our case here, it is described by spiritualists among the resistance's rank and file: revenge killings – or, as Hinton calls it, 'disproportionate revenge'. Commanders deemed to be overusing their powers in front of their soldiers and thereby humiliating them may have faced revenge someday:

> If a commander overused his power during battle [e.g., humiliating his subordinates], he would be in trouble. If soldiers hated him, they could simply kill him during battle. Sometimes the commander used his power too much such as using the gun the knock on a soldier's head. Soldiers then might remain silent and say nothing for a while, but eventually they might secretly kill their commander in battle. Commanders who died in combat were usually killed by their own soldiers. (RF-F1)

A knock on a soldier's head was answered by murder after a long time of silence and feigned compliance. Accumulated humiliations were answered by drastic means that appear to be disproportionate when comparing single events (knock on a soldier's head) with their means of retaliation (murder during combat). There were two ways to deal with humiliations by powerful commanders: Soldiers could either believe that cosmic powers would retaliate and kill the commander by, for example, causing him to accidentally step on a mine. Or soldiers could retaliate for the breech of spiritual rules of conduct – by secretly killing the commander during missions far away. Oftentimes these two possibilities appear to be chronological in nature, as the soldiers trusted in cosmic retaliation at first and only took action if this appeared to fail.

Blank-Page Warriors

It is debatable whether respondents from the group of blank-page warriors differ from blank-page operators. All that separates them in habitus formation is their position in the field and the fact that those in the mid-range made a modest career within the military apparatus. For many, it seems, this upwards mobility was made possible or at least supported by certain social resources within the field. Blank-page warriors had connections to people who introduced them to the party (this was largely a prerequisite for joining anyway), but most of these individuals were rather low-ranking combatants and commanders. Another reason may be that they did not prove to be as 'active' and therefore suitable for promotion as others did. Some reached a comparatively high level in command at the very end of the war (battalion commander at best), but this happened at a time when the Khmer Rouge had nothing else at hand with which to reward their loyalty. Yet another reason why they did not make a career might be the fact that it was rather risky to rise in rank, due to the fact that it might draw Angkar's attention during internal screenings for hidden spies and counterrevolutionaries. With regard to Jeremy Weinstein's activist and opportunist divide, such individuals posed yet another theoretical problem: Their idealist stance did not result in but was the result of participation in the Khmer Rouge military. It was the result of being part of a disciplinary structure for many decades, that is, of indoctrination, moral and political training, separation from families, and close-knit disciplinary and pastoral control.

Habitus Formation and Lifecourse

The background of the blank-page warriors was largely the same as their mid-range commanders. Like their direct superiors, they all came from rather impoverished peasant families, and they received a comparatively low level of education, if any, as many of them were entirely illiterate. Some of the respondents also mentioned the loss of a parent at an early age (e.g., at the age of 7: BA-KR2) and that they did not remember much about their life within their families (e.g., RF-KR1). Most of them studied at best for two or three years until grade 10 (back then, the grade count ran backwards starting with grade 12): "I received little education. I quit school when I was in grade 10" (RF-M1). After dropping out of school, some went to a pagoda as a novice monk in order to carry on with their studies or simply to have a place to live and eat. However, most joined the Khmer Rouge after a short while, either due to the US bombings, severe poverty, Sihanouk's call to bring him back to power, or conscription by the Khmer Rouge military.

Access and Mobility

There were several different stages of recruitment within the Khmer Rouge military and security apparatus. At the very beginning, recruitment largely failed due to 'analytical errors' on the part of primarily the intellectual and middle-class leadership. They failed to realize that some areas were poorer than others because they assumed that the worst conditions they had seen in certain areas were representative of the whole countryside, leading to an incorrect assessment of a potential pool for recruits among the agrarian populace (cf. Willmott, 1981). Beyond anti-Colonial fighters who had served for ages and the upper-class leadership, only a few of those who were considered to be the 'backbone' of the revolution joined their ranks. However, two events accelerated recruitment during the early years. The first and widely discussed event was the US bombing of Cambodian territory in an attempt to destroy bases of the Vietnamese communists and their support lines via the Ho Chi Minh trail, which ran through the remote parts of eastern Cambodia. The scope of ordnance being dropped over Cambodia was revealed at a rather late point in time, when US President Clinton declassified a number of documents during his trip to Vietnam in 2000. According to these documents, the bombings started as early as 1965 (four years earlier than previously estimated) and ended in 1973, with 2,756,941 tons of ordnance being dropped in 230,516 sorties on 113,716 sites. And as Kiernan and Owen pointed out: "Just over 10 per cent of this bombing was indiscriminate, with 3,580 of the sites listed as having 'unknown' targets and another 8,238 sites having no target listed at all" (Kiernan & Owen, 2006, p. 63). This bombing campaign, which during its peak stage was named 'Operation Menu' (including air raids correspondingly nicknamed Lunch,

Snack, Dinner, Supper, and Dessert), was designed to destroy the heterogeneous Viet Cong insurgency and ran through large parts of Cambodian and Laotian territory with devastating social, political, and economic effects.

While the bombings started under King Sihanouk's *Sangkum Reastr Niyum*, they widened in scope after the pro-US government under Lon Nol and Sirik Matak toppled the king from power in 1970. The actual impact this had on recruitment by the Khmer Rouge is still a matter of debate. The economic impact for many was certainly unbearably high: "Many Cambodian farmers incurred much economic difficulty after having massive damage done to their farms. Re-constructing efforts were just too costly. Many were left with heavy financial debts, having to sell off their property to pay their debts off" (Procknow, 2011, p. 109). Pol Pot himself propagated that as many as 600,000 deaths were directly caused by these air raids – a number that remained widely in use for many years if not decades, sometimes increasing to 800,000 or even 1.4 million Cambodian victims. In comparison, Henri Locard cites Marek Sliwinski's estimate of 240,000 victims, which would include all deaths during that period, meaning also those caused by the Khmer Rouge themselves (Locard, 2000, p. 12). Others suggest the number 150,000 plus thousands of displaced people, which at least made a certain recruitment strategy possible, yielding an increase in recruitment for the Communist top command. Combined with propaganda putting the blame for impoverishment and misery on the US 'lackeys' in Phnom Penh, local cadres were able to recruit mostly young adults within villages:

> Party cadres would go around informing the people that the Lon Nol government was to blame, that Lon Nol himself requested that the Americans bombard the Cambodian countryside, responsible for the devastation and suffering of innocent agrarian villagers. They would entice people to join the revolutionary army as the only way to stop this destruction. (Procknow, 2011, p. 110)

A Khmer Rouge officer highlighted how easy it was to recruit the children of villagers after bombs fell on their houses and rice fields, making them wander around mute for several days:

> Every time after there had been bombing, they would take the people to see the craters, to see how big and deep the craters were, to see how the earth had been gouged out and scorched ... The ordinary people sometimes literally shit in their pants when the big bombs and shells came. Their minds just froze up and they would wander around mute for three or four days. Terrified and half crazy, the people were ready to believe what they were told. It was because of their dissatisfaction with the bombing that they kept on co-operating with the Khmer Rouge, joining up with the Khmer Rouge, sending their children off to go with them ... Sometimes the bombs fell and hit little children, and their fathers would be all for the Khmer Rouge. (Kiernan & Owen, 2006, pp. 67–8)

Among long-serving Khmer Rouge, these bombings were often cited as reason for joining the Khmer Rouge in the first place, connected with a considerable feeling of anger that made them volunteer: "In late 1971 [an] aircraft bombed my house and killed my mother. In early 1972 subdistrict cadres came in to recruit for the army. Because of my pain over the bombing, I volunteered to join" (Huy, 2003, p. 11). The Khmer Rouge cadres coming to villages held the Lon Nol government responsible for all the hardships and bombs that befell the agrarian population, destroying their homes and fields or even killing members of their family. Toppling the Lon Nol leadership was presented as the solution to all miseries. However, it is still difficult to assess the actual impact of bombings on rates of recruitment within the Khmer Rouge. It seems, as also highlighted by Kiernan (1996, p. 7), that the bombings led to an increase in recruitment and presented a viable strategy for the Communists to build up their forces but nevertheless was not sufficient to seriously threaten the incumbent government.

Aside from that, bombings started as early as 1965, but it was not until 1970 that the Khmer Rouge were able to recruit a stronger force. Most decisive seems to be the call via radiobroadcast made by King Sihanouk after being ousted from office in 1970 to join his forces in the 'marquis'. Sihanouk formed a coalition with the Khmer Rouge, who, as he believed, would be an effective force to bring him back to power. In the end, the Khmer Rouge only became an effective force precisely because of his call. Most of the respondents interviewed in this project, up to those in mid-range positions, mention that call and their will to restore the royalist (not Socialist) order as their main motive for joining the Khmer Rouge. Sihanouk boosted the Khmer Rouge's credibility and legitimacy, strengthening their military apparatus. This made them strong enough to capture certain 'liberated zones' all over the country but still not strong enough to capture the capital. This became possible only after the withdrawal of the US military from Southeast Asia in 1973, which left a power vacuum behind of which the Khmer Rouge could make use with its boosted force.

Starting in 1973, the Khmer Rouge also changed their main strategy from volunteer recruitment to village-based conscription in 'liberated areas'. Local cadres selected at least one son or daughter from each family to serve in the revolutionary movement, mostly in a mobile youth brigade to start with (e.g., CC-KR2). One respondent said that it was simply 'normal' to join, since everyone else did as well (BA-KR2).

During the years of the Khmer reign, recruitment was made even easier, since families were torn apart and children were forced to live in areas far from home. Families had no rights over their own children who were put into indoctrination camps, oftentimes tortured, and their revolutionary spirit 'tested'. Only the good and pure would survive tests and could be eligible to join core organizations such as the military and security apparatus. While their background had to be pure, their minds and spirit had to be tested as well. Pol Pot himself described his recruits as "completely illiterate people who did not have even the slightest idea of cities, automobiles and parliaments, but who dared to fight under the guidance of the party" (Procknow, 2011, p. 114). Class and political purity were prerequisites for recruitment. But in many lower administrative positions, a certain social background may have been enough, opening doors for rather impure if not formerly criminal cadres:

Another problem with the regime was the arbitrary reality of appointments to local positions of power. In practice, the line calling for appointment of poor peasants as village, cooperative and subdistrict cadre meant that anyone of that class background was appointed, even if the reason for their poverty was that they were hooligans, people who had sold off their family inheritance to pay off gambling debts. Once in power, the ignorant poor – having no idea how to make a proper living – were even more oppressive than their predecessors. (Interview with Van Rith, 2003)

However, for members of the military in particular, the main mode of recruitment was to select cadres according to a twofold eligibility. In addition to coming from a pure class, a recruit had to be 'absolute'. A former member of the military described how he was selected as an 'absolute':

They were looking for the 'absolutes;' I did not understand what it was. [Translator's Note: 'Absolute' was used to describe those who were especially daring and dedicated.] When they looked for 'absolutes', I raised my hand and I was picked. I was made a soldier. I was to be an 'absolute' soldier. […] I raised my hand because I saw others do so. (Interview with Lat Suoy, 2011)

To what extent a recruit could be trusted as an 'absolute' was not just a matter of assessment during initial recruitment but also during constant tests at the beginning (e.g., going to battle without a weapon but only a pan to cook rice), constant re-writing of autobiographies to fit the party line, constant livelihood meetings to build yourself, and close-knit surveillance in daily activities, in which no one was allowed to do anything without the commander's approval. Only those who proved absolute in spirit and eager to fulfill their tasks could hope for promotion: "For the ones who demonstrated hard work, the Khmer cadres would note this and recommend these selected few to mid-cadre level leaders that certain workers should be considered for promotion to the party leadership ranks" (Procknow, 2011, p. 141). Promotion was the only possible reward. But getting promoted could mean getting better rations of food and more rights (e.g., to move, to see your family, etc.). However, it was also a double-edged sword since those who were positioned higher within the chain of command often fell victim to internal purges as well.

The situation changed a bit after 1979 and even more after 1993. At times when the Khmer Rouge faced high levels of defection, it seems that they started to recruit more hastily without major initial biographical reviews (cf. Heder, 1996). But still, most recruits had been rather young, and a certain social background remained favored. In general, none of the respondents within the lower ranks had decisive social resources that could have brokered them a position in the mid-range leadership. This, and maybe a certain lack of 'revolutionary spirit', seems to be why they remained at the bottom of the hierarchy. While the lowest level was selected along twofold eligibility, upper ranks largely seem to reconstitute certain patrimonial networks as well, mixing the pure and absolute with the well-connected. While vertical mobility was easier for agents with social resources within the field, recruitment and testing took rather long for the rest, who also stayed at the bottom until the very end of engagement within the Khmer Rouge apparatus: "It took me nine months to become an official core member of the faction" (RF-M1). After joining at an age of twelve years and older and going through eligibility tests, the rank-and-file soldiers largely "graduated into combat" with "experience as their teacher" (Interview with Van Rith, 2003).

Classificatory Discourse

Combatants within the Khmer Rouge rank and file experienced a symbolic increase in standing and gains in power, but in the end, the status of respondents in this habitus group within the field remained at the bottom of the internal hierarchy. Recruits within the security apparatus were praised as the vanguard of the revolution, and the chances of increasing one's status within the apparatus after proving a proper spirit were comparatively high. Therefore, recruits could hope for certain fame and powers connected with the twisted valuation of the peasantry and youth by the top command of the Khmer Rouge. As one respondent highlighted: "In the military, people could get positions, fame, and power" (CC-KR2). Positions and power over people they could not have hoped for before the Khmer Rouge came to increase their status. Cadres, particularly those in the security apparatus, now came into the position of commanding those former 'exploiters' who had been beyond their social reach within the 'old order'.

What characterized the discourse of the Khmer Rouge rank-and-file soldiers was the ambiguity of being an ideologically firm and empowered backbone of the collectivist revolution and simultaneously being in constant danger that may even lead to an increase in powers and 'fame'. No one was to become too visible, but they sought to be noticed enough to be promoted to the core as an 'active' revolutionary. In the end, respondents within this group lingered at the lowest rank-and-file level without receiving higher positions and moving up the chain, or they experienced only slight improvements up to the rank of a company commander. One reason for this is certainly their lack of social resources. Not a single combatant mentioned having contacts beyond those who introduced him to party membership. These contacts had been previous members of the military as a core organization, but at comparatively low levels. Combatants interviewed at this level were situated outside of patrimonial networks seen within the mid-range and top command of the military. Their main and sole resources were their class background as well as their proven loyalty by devotion to Angkar's regulation and 'clear-sighted' plan.

The rank-and-file soldiers incorporated large parts of the Khmer Rouge discourse into their behavior, specifically keywords such as selfless devotion, practice, and collectivism ('sameness'). As former child soldiers, they lived within the Khmer Rouge apparatus for more than twenty years and still largely live in Khmer Rouge dominated villages. For them, good morality meant being close to their subordinates and to respect the command: "I was always sticking to good morality and always kept present with my soldiers over the day. My policies had been proper and I respected my commanders. […] I always stayed with my soldiers" (CC-KR2). A good soldier had to constantly be mentally 'active' while executing his tasks:

R: A good soldier must be active both in attitude and wisdom.

I: What do you mean by active?

R: Active meant working without getting sick of it. We always appeared in all the three-to-four battles per week. This was called being active. Inactive meant that we appeared in only two battles and were absent during the other two battles of the week. (RF-KR2)

A good revolutionary combatant always sacrificed himself for the sake of the collective, that is, the nation and its incorporation, the party. Skills were seen not as an individual capacity but solely the capacity necessary to get rid of individualism. Demanding something for yourself might have been considered treachery:

I: What did you do make yourself become a skillful soldier?

R: That relied mainly on our nationalistic stance. We were fearful that we could be accused of being a traitor. That was why we served in the military without caring about any salary. Everything we did was for the sake of our country. (BA-KR2)

Thus, a good life became possible when a soldier adopted a nationalist mentality: "Good life means to have a nationalistic mentality" (CC-KR2). Concretely, this meant that they had to sacrifice themselves for their commanders, the representatives of the collective: "If we didn't love our commander, who should we love instead? We always followed him. If he was alive, we would be alive. If he died, we would also die" (CC-KR1). Love your collective, love your commander, and avoid being distracted by individual emotions, such as those you feel for your family. A brave and clear-minded soldier, therefore, had no family and no wife:

I: How to become a brave soldier?

R: We did nothing special. Those having no family or wife were clear-minded. They didn't care about anything else than [being a good soldier]. (RF-KR2)

Within collectivism as the rank-and-file soldier understood it, no one was different than others, and no one was to have individual pleasures. Thus, commanders and soldiers were all the same:

I: How was the personality of the Cambodian soldiers back then?

R: They were all the same. (CC-KR1)

Not just soldiers but also commanders were basically the same. No one deviated from the line, since everyone came from the same school: "Commanders were not much different actually. They would have different leadership styles only if they had gone to different schools. But they all went to the same school" (RF-KR1). No one would even have dared to say that he was special in any respect, as this might imply a qualitative difference and imply the superior conduct of an individual:

I: How was your leadership style different from others do you think?

R: I dare not conclude that my leadership was different from others. We can't simply say that our leadership style was better than others'. (BA-KR2)

People within the collective stayed close to their comrades. Everyone only did what he was assigned to do, i.e., stay with his comrades and not enjoy personal benefits:

The most important thing was that I always stayed close the soldiers, and that I didn't always command them inconsiderately. When we commanded soldiers, we only had to do that. Commander's role at that time was to stay directly with the public mass, and not to just sleep and eat comfortably. (BA-KR2)

Thus, the party and those in the lower ranks presented themselves as not enjoying any special treatment or benefits. However, the upper ranks did enjoy benefits such as better food rations, and the rank and file benefitted from being in a position of power over 'new people' and other civilians.

Akin to the blank-page leaders, everyone in the Khmer Rouge rank and file highlighted the primacy of practice over theory. Again, the analogy of learning a language was cited as an example: "The more we fought, the more we learned. It was just like how we learned a language" (BA-KR2). Most important was whether someone was fit in not theory but in practiced bravery and proven strength: "During the Khmer Rouge regime, they only selected skillful people and those having strong fighting mentality. It was not important if a person was bad at linguistics or theory. They mainly chose strong fighters" (CC-KR1). Similar to others in the field, proven bravery in practice was the main symbolic resource within these ranks. Being literate was not a criterion for promotion (actually it was quite the opposite), but strength and bravery in fulfilling soldierly tasks were rewarded and recognized: "A strong soldier was the one who was brave and who followed orders. This kind of person could become a commander. Even if we knew English, French or whatever language, we wouldn't be selected because they only chose brave soldiers" (CC-KR1).

The idea of democratic centralism, that is, the handing down of the 'correct' and 'clear-sighted' plan to the subalterns of Angkar, also dominated the rank and file. All respondents highlighted the hierarchical nature of Khmer society and the high level of respect for discipline as shown by soldiers: "Our Khmer soldiers were special in the sense that we respected discipline. They were organized hierarchically" (CC-KR1). Being disciplined and respecting the plan created by the party's center was essential: "We had to respect the plan and discipline, and be gentle. Then they would respect us" (CC-KR2). Soldierly skills were not viewed as something individual within a collective, in which the correct plan was handed down and its implementation was a matter of soldier's properly carrying out their tasks. To become skillful meant to be flexible, active, and hard-working when implementing plans or commands. Skills were the sheer result of following orders:

I: How can you become a skillful soldier?

R: I followed the commander's commands and I always performed them well. [...] You had to be flexible and follow orders. (CC-KR2)

The rank-and-file soldiers incorporated major pillars of the Khmer Rouge discourse into their actions, including collectivism and sacrifice, the primacy of practical experience, and the principle of democratic centralism as obedience to the clear-sighted plan that was handed down by the collective's center.

Khmer Rouge rank-and-file soldiers exhibited the same ethical guidance pattern of increasing the value placed on their acts as soldiers as did the rank and file within the non-Communist factions. The patterns are all the same. The act of killing a Vietnamese was seen as a must, but killing Khmer was viewed a sin; thus, preventing their deaths even elevated a combatant in his moral superiority. An attack could not be prevented, but a soldier could decide whether to fight (seriously) or not: "Whether I attacked or not depended on me. When I saw Khmer civilians, I wouldn't fight. But if I saw Youn, I would" (CC-KR1). This soldier even gave a concrete example of an operation in Samlot, near Battambang. Here, he decided not to detonate an already prepared mine because 'K5 people' showed up at the scene, that is, Khmers who had been recruited by force to construct a defense belt to seal off the border:

> Earlier on, I went for an operation in Samlot. I already prepared all the mines. But then K5 people were carrying weapons and food for Youn soldiers. I saw that there were actually only a few Youn soldiers, and the rest were all Khmer. So I decided not to detonate these mines. I had a similar experience later on. I saw civilians walking, so I didn't explode the mines. I thought only Youn would travel there every day. But when I went there, I only saw civilians walking. If I had let these mines detonate, there would only be death and sin. If they were all Youn, I would definitely do it because we were only against Youn. (CC-KR1)

Power Practice

The whole disciplinary apparatus of the Khmer Rouge was tight-knit, leaving only few if any possibilities to evade control. Some say in retrospect that they had been 'hot-tempered' and willing to fight: "I was young at that time, so I was not afraid of death. When I was told to fight, I would do it" (RF-M2). Others, by contrast, felt fooled, noting that they joined the Khmer Rouge with childish expectations, only to be disappointed by reality: "I saw other people having fun being a soldier, yet it was not how it was like in reality [when I became one myself] as there were a lot of fights" (RF-M1). Every rank-and-file soldier reported that he was immediately sent to battle: "I did learn practically during the battle. I didn't go to any school for it. I simply learned directly from battle experience" (CC-KR2). But while training was largely skipped, discipline was constantly and closely monitored. Even the slightest sign of disobedience was punished severely, and many things were

6

considered disobedience, including simply starting "to argue with us" (CC-KR2). As in any area of the Khmer Rouge apparatus, the military was clearly regulated and there was no space for 'arguments'. Because of the collectivist system, there were no rewards for soldiers or commanders, no matter what they did (RF-M1). Everything was distributed equally within the units:

> [Our commander] didn't give us anything during then, but he gave us food. If there were noodles, everyone would get noodles. If there were sardines, everyone would get sardines. If there was some fish, everyone would have fish. We all got things equally. (CC-KR2)

Military collectives were as regulated like any collective under the Khmer Rouge. Regulations were passed down in great detail, affecting most private spheres of the soldiers' lives. For instance, marriage was considered a threat to the collective and the fighting spirit of the individual, therefore commanders were in full control of if and when a wedding would take place:

> To be frank, we didn't follow Khmer traditions here. We never betrayed our commander though. For example, you are my commander. I told my commander that I loved a person. But because the commander was afraid that I would not work hard for the movement, he didn't allow the marriage yet. (RF-KR3)

The commander had to take care of the 'livelihood' of his combatants, which included all of their actions and thoughts. Most had to write biographies during recruitment and constantly thereafter. Criticism sessions or 'livelihood meetings' to assess strengths and weaknesses took place on a weekly basis as well as before and after combat operations: "They would criticize us. The misbehaved soldiers would talk about their own mistakes, then they would subsequently be criticized" (RF-M1). During morality trainings and hours of political indoctrination, the commanders advised their subordinates on how to have a clean livelihood or 'living morality': "The senior commanders talked about living morality. They basically spoke about clean morality" (CC-KR2). Criticism sessions were to 'alert' members of the collective on their daily mistakes, and even soldiers were at least formally allowed to criticize their superiors if they dared to:

> We told him [of his mistakes] because we wanted to alert him [our commander]. No one could be completely good. It was normal to make some mistakes. We must help explain each other. When my commander did something wrong, I instructed him back. At that time, we normally had a self-criticism meeting. The commanders could welcome their friends or soldiers' comments. (CC-KR2)

Within the disciplinary system, no one was safe from collective surveillance. Most of the respondents highlighted that these criticism sessions were sufficient for solving problems related to discipline after 1979.

In case of indiscipline towards the collective, soldiers were to be 'educated' or rectified continuously:

> If someone didn't follow orders, we would try to educate them and be good to them. For example, if they didn't follow us, we tried to educate them again and again. Sooner or later, they would follow. It was important to educate them continuously. (CC-KR2)

Since good revolutionaries were seen as those who were active, it was laziness which was seen as the source of disobedience: "The source may have been laziness. They might find the fighting too complicated, so they didn't follow" (CC-KR2). Daily reports from unit heads to superior cadres on the situation within their unit and the activities of their subordinates also helped to maintain an atmosphere of constant surveillance. Most soldiers started to incorporate the ideology and control into their thoughts and actions, thereby rectifying themselves to appear as good revolutionaries in the eyes of Angkar.

However, it was clearly not just a system of constant surveillance, of constant indoctrination from childhood onwards, and of thousands of tiny pinpricks within livelihood sessions that made the rank-and-file soldiers obey and adapt to self-rectification. Punishments for misbehavior of any sort were extremely disproportionate. Combat control was particularly fierce. Soldiers who ran away due to fear faced immediate response from the group commander awaiting them in the back: "If they ran to the back, the commander would shoot them. Khmer Rouge were so extreme" (RF-M2). Another rank-and-file soldier responded to the question of what commanders did to convince them to fight: "Khmer Rouge said nothing. They simply shot those who ran away from combat" (RF-M1). While higher commanders used civilian intelligence and T.O. communications to check on their units in action, individual soldiers were monitored by their respective group commanders: "It was possible to check up on [individual soldiers] as well. There was a commander in each group, so no one could escape" (CC-KR2). However, not everyone was shot right away, but many had been 'alerted' with regard to their 'obligations': "We would explain the situation to them. Then they would no longer be afraid of battles. We instructed them about their responsibilities in combat, the reasons to go there and our obligations" (CC-KR2). In any case, 'free movement' was considered selfish and individualistic. In addition to bans on individual trade, gardening, marriage, and 'living morality', leaving a barrack without permission was met with physical instructions and re-education: "Freely leaving the barrack without asking for permission [was considered bad]. If people were too free, they would be hit" (RF-M1).

Rectification of the Self and (the Lack of) Lines of Flight

Soldiers of the Khmer Rouge security apparatus and the rank and file had been rectified and indoctrinated since their early youth, most living within the party for almost three decades and staying close to their former comrades in villages along the border thereafter. All highlighted that they always obeyed the command and did not think about deviating from the line. Almost no instance of refusing orders was raised, and most even stated that it was rather unimaginable for them to do so. Although deviation from Khmer Rouge policy would lift their moral status in front of a foreigner and a Cambodian student, during a time when all of the horrors under the Khmer Rouge become widely known and legally pursued, those living along the border in Anlong Veng or Pailin did not deviate from Khmer policy. Some may have feared retaliation by their former comrades; others were simply ideological hardliners after decades of indoctrination and rectification. And, of course, those still living in Khmer Rouge dominated areas spoke rather enthusiastically about methods of rectification, such as livelihood meetings, for instance, while those who defected and escaped to the interior delivered more drastic accounts. Certainly the combination of constant rectification with a highly disproportionate disciplinary system contributed to this lack of lines of flight.

Within the interviews conducted for this project, there were only few sequences on rectification and lines of flight. Everyone had to deal with his fear on his own, while showing signs of weakness would threaten his candidacy as an 'absolute'. Learning by doing and adapting to the situation was mainly a matter of time and routine. Watching the behavior of others could serve as practical training manual: "I could learn from practical experience after some time. At first, we had to observe others on how they run, and how they escape, and just follow their example" (RF-M1). The only means of self-rectification within the Khmer Rouge rank and file was routinization and seeking some protection – not spiritual but practical means of protection:

> R: I had to solve it [my fear of combat] by myself.
>
> I: How did you do it?
>
> R: I spent the first case of bullets to shoot randomly. And when I saw other soldiers running, I would follow them. I also sought for small hills and trees [to hide]. (RF-M1)

Religion was regarded as reactionist and was officially banned. Questions regarding spiritual protection, therefore, largely made no sense to a Khmer Rouge combatant. All they had was the proverbial milk from their mother, which would lower their fears in combat:

> I: Did you receive any Yorn or something else from monks?
>
> R: No, we always thought of our mother's breast milk [when we were in danger]. (RF-KR2)

Respondents within the rank and file showed a similar 'opportunistic' attitude like the pragmatists and, of course, the blank-page operators, highlighting that they did not choose to fight for a certain faction. Much to the contrary, they simply served whoever was on top: "Back then, we didn't know which side was right. We served for different political factions. When we were there [under the Khmer Rouge], we served them. When we came here [under the Hun Sen government], we serve them" (CC-KR2). The only thing that was clear was that a Vietnamese was worth wasting bullets on. Their eagerness to kill Vietnamese was only hampered by a policy introduced to save some bullets at times of scarcity: "In 1979 when we fought with Vietnamese, we never saved our bullets. But later, we had a policy of saving our bullets. It was known as 'one bullet, one person'. One bullet shot must kill a Vietnamese" (CC-KR2).

Due to the close-knit system of control, only few chances arose to prevent engagement in combat. Lat Suoy provided an instance of how he tried to evade combat – an instance that also proves how difficult it was, ultimately leaving only the tiny possibility of maiming yourself:

> I was trying to shoot myself in the lap in a way that would not hurt the bone. I checked myself over. The bullet could hit the flesh, but not the bone because it could cause amputation. I shot myself in the lap like this. The bullet did not get through my flesh. It did not hurt me. Therefore, I thought to myself that I had to find another way to hurt myself. I starved myself for three days. (Interview with Lat Suoy, 2011)

However, if they chose to maim themselves, they had to do it well, because Khmer Rouge had medics to monitor their health and sort out those with fake injuries and illnesses: "Some lied to us. They told us they were sick although they were actually not. We also had our medical staff, so we asked the staff to go check and we eventually found that those soldiers were just fine" (BA-KR2). Defection was seldom an option, simply due to the fear of retaliation upon a soldier's return to the interior. It took more than a decade until morals broke down, war fatigue set in, and demobilization became possible, providing a window for many combatants and commanders to slip through and end decades of involvement in combat.

Chapter 7
Sociology, Civil Wars, and Conflict

There are a couple of deviations from Pierre Bourdieu's approach within this study that are needed in order to make use of the previous insights 1) on the sociological analysis of civil wars and conflict, and 2) on the analysis of social structures in general and outside of Europe and the US in particular. Many of them relate to Bourdieu's concept of social space, but, for the most part, they may be resolved – with some slight changes – using his concept of a 'field'. The first point of deviation is that the current study places a stronger emphasis on the lifecourse of agents as a diachronic factor that accounts for differences in habitus formation. While Pierre Bourdieu highlights diachronic aspects of habitus formation in his theory, he mainly uses statistical methods (correspondence analysis, in particular) to detect differences between individuals that matter for their habitus formation. Within his seminal study 'Distinction' (2010), variations in habitus formation within the French social structure are analyzed through the construction of a two-dimensional social space. Using correspondence analysis, Bourdieu structures a society's social space along two dimensions defined by economic and cultural resources on its x-axis and y-axis, respectively. Depending on the number of resources at an agent's disposal, their composition, and their current tendency (e.g., increasing/decreasing) at a given moment in time, his position on the table and the corresponding habitus can be determined alongside likely lifestyle patterns. According to Bourdieu, positions within this social space are homolog, that is, the distance between agents in all fields of practice remains the same. The social distance on the flat space of the social structure becomes reiterated within all social practices across all fields and even inscribed into physical space (Bourdieu, 1996). It is a nation's current space of social structure that matters most for the formation of the habitus, not participation in different fields during the lifecourse of the individuals. By using a snap-reading method such as statistics – specifically correspondence analysis, in this case – Bourdieu tends to overemphasize synchronic aspects of habitus formation and also binds them to unambiguous positions within a nation's social structure. However, as Bourdieu himself constantly highlights and also tries to make use of using additional qualitative methodologies such as more or less structured interviews, "The social world is accumulated history" as well (1986, p. 241).

At most, however, that history is understood as an agent's self-adjustment to a preset life trajectory inherent to his social position. Bourdieu calls this mechanism 'social ageing' in which people "become what they are and make do with what they have" (Bourdieu, 2010, p. 105). By and large, the volume and composition of resources inherited within the family's milieu prescribe which biographical lifecourse is probable for a given individual. Socialization would be a redundant concept, although Bourdieu stresses that his neglect of socialization theory was simply due to time constraints (e.g., 1997). However, on occasion, Bourdieu also highlights a dialectic between a prefiguration by inherited capital and either collective or individual events that may change the socially determined course of fate. Thereby he plays upon a dialectics between structure and events which are not just part of the habitus' spontaneous intervention:

> To a given volume of inherited capital there corresponds a band of more or less equally probable trajectories leading to more or less equivalent positions (this is the *field of the possibles* objectively offered to a given agent), and the shift from one trajectory to another often depends on collective events – wars, crisis, etc. – or individual events – encounters, affairs, benefactors etc. – which are usually described as (fortunate or unfortunate) accidents, although they themselves depend statistically on the position and disposition of those whom they befall (e.g., the skill in operating 'connections' which enables the holders of high social capital to preserve or increase this capital), when, that is, they are not deliberately contrived by institutions (clubs, family reunions, old-boys' or alumni associations etc.) or by the 'spontaneous' intervention of individuals or groups. (Bourdieu, 2010, p. 104)

Hence, you are not simply becoming what you are. Although the social background of an agent's family presets important trails within an agent's lifecourse, there are still 'events' – individual or collective – that may to a certain degree (and maybe depending on the agent's age) shift its direction and herewith the agent's habitus alongside his or her acquired resources and schemes. The problem, however, remains that Bourdieu largely used statistics as his primary analysis method and by and large resorted to interviews for purposes of illustration rather than as a means of detecting differences in the ways the lifecourse shifts an agent's habitus formation in comparison to others. Moreover, Bourdieu's reference to wars and crises is largely abstract and refers to rather drastic changes (although of course this suits the framework of this study quite well). As long as he refers to a fixed position on a social space that is homolog on all fields of practices, shifts remain rather mysterious.

This study, however, found lifecourse information that was decisive for the *formation of differences* between habitus groups. These could sufficiently be explained neither by the social background of the individuals' family of origin nor by their position in the Cambodian society or the field of resistance. At the same time, there was a clear unifying biographical pattern underlying the difference that cannot be explained by the creational aspect of the habitus as merely a 'structuring structure'. Hence an outline of social differentiation also needs to take into account the individuals' lifecourse and its intersection with historical and social change. By doing so, differences in habitus formation will come to the fore that otherwise would have gone unnoticed or would have remained impossible to explain. For example, differences in habitus formation within the rank-and-file soldiers of the movement are not possible to detect using solely synchronic methods. They all came from a similar milieu in which their parents were peasants who lived on subsistence farming. They all had been low-ranking soldiers within the field, owning barely more than a weapon and being fed by aid distributions and their commander in charge. All accessed the field by enlisting as a rank-and-file soldier without previous social contacts within the resistance. Moreover, they did not strive for or reach a higher position within the field. There was no competition to improve their position but solely a drive for sufficient food, safety, and respect (which, by the way, contradicts Bourdieu's assumption that all social fields are characterized by omnipresent competition). All of the rank-and-file soldiers would classify themselves as sons of 'ordinary' lower-class peasants who became 'ordinary' lower-rank soldiers simply due to conditions in times of war that displaced them from their home communities in search for food and security. Yet, still, their habitus is very different. As shown within the previous chapter, their habitus mainly differs because of variations within their lifecourses that first and foremost relate to different experiences of war at certain points in their personal history or biography (fitting Bourdieu's description of an 'event').

This social position, however, is not primarily one within the Cambodian state and its social structure. This brings us to the second point of deviation. The field of insurgency defies being located in the Cambodian nation state and its social structure not only due to its geographic location and an overriding influence by international powers but also due to the transnational biography of the people within and the ambiguity inscribed in the resources they managed to acquire during their lifecourse. While Bourdieu's concept of fields may be used to analyze social figurations and agents that are not unambiguously situated within a state, his analysis always takes the nation state as a point of reference and excludes transnational institutions, agents, and forces (Rehbein, 2003). For example, returnees from exile, such as former members of the Cambodian elite who spent years abroad while being deprived of their citizenship, defy being unambiguously positioned within the Cambodian social structure, especially while living in refugee camps situated in Thailand. The same applies to former soldiers who already lived in Thailand for many years and worked on rice fields for Thai farmers or as intelligence personnel for the border police. Large parts of their habitus were formed abroad, and they acquired many of their resources in other countries; thus, their value is at least contested upon their return to Cambodia. Many of these 'newcomers', however, never actually entered Cambodia to live there and later on returned to the Diaspora communities in either France or the US.

Hence, not just transnational migration and figurations pose a problem but also the question of how to quantify the acquired resources and their value within the field. While some refugees brought along some gold or managed to gain control over certain lucrative flows of goods and people, the value of economic means in a society at war is simultaneously high due to its scarcity and low due to the omnipresent threat of being gunned down in a robbery, for instance. Moreover, most of the economic resources are allocated by physical force (Schlichte, 2003; 2004). Therefore, many resources have a rather erratic and unstable value within conflict. During war and in the closed and crowded space of a refugee camp, a weapon becomes one of the most valuable resources, oftentimes the deciding factor when it comes to the allocation of basic goods. But is it an economic resource?

Is it a cultural resource if people know how to handle it? Or, is it symbolic due to its symbolic 'force' and the prestige connected with it? Similarly affected within the transnational field at war are acquired vocational skills or certain areas of knowledge, while social resources, by contrast, increase in importance if not becoming the most decisive resources (see also Schlichte, 2004). Within war, many resources become highly volatile: A person's reputation of invulnerability and strength, for instance, may cease almost immediately or increase after if it is widely noticed and 'gossiped' about in a combat operation.

Furthermore, in order to grapple with the social figuration of the field, more weight needs to be placed on social and symbolic resources than Bourdieu acknowledges. Thereby the field position of certain groups whose status rested upon more than solely economic and cultural resources becomes easier to understand. As mentioned before, Bourdieu focuses on economic and cultural resources to erect a nation's social space and detect differences in habitus formation. The two dimensions of social space within the diagrams he uses are defined by the amount and composition of economic and cultural resources of which agents dispose. Their habitus corresponds with the related position within a nation's social space. Thereby the possible differentiating influence of social or symbolic resources on the positions of agents within society and the corresponding deviation in habitus is excluded. A two-dimensional space, however, is not sufficient for explaining the positions of agents within different fields. While cultural and economic resources may be most decisive for the habitus formation within the French society he was studying (especially when sorting agents along occupational lines and thereby excluding many others), social and symbolic resources are at least similarly important in understanding differences in habitus formation and social status within our field of study. Moreover, both resources have certain important qualities and characteristics due to the particular history, set-up, and relative position of the field of insurgency. These are also difficult to quantify statistically (e.g., the symbolic value of surviving a battle or having bullet scars) and therefore call for a qualitative approach as well.

Moreover, in contrast to a nation's social space with unambiguously assigned positions, an agent's position within a field is always relative to other fields and only one of many of which he can be part. The flat structure dissolves into various fields that stand in relation to each other but are nevertheless autonomous. This makes multiple simultaneous social positions possible, which may also change due to changes in the relationships between these fields. For example, when the resistance gains strength, the influence of leading agents on other fields rises as well. Another strength of Bourdieu's field concept is that various types and different values of resources can be relevant, depending on the rules of the respective field. It is a matter of research to establish which resources matter and how and to develop the specific rules of their valuation within the respective field. Thereby, for example, we can understand why agents who came from long-time influential families within Cambodian politics can have the same level of influence as those who fought many battles as rank-and-file soldiers, even though they previously had neither economic, social, cultural, nor symbolic resources at their disposal. The positions that agents take are not fixed but are always relative to the particular fields to which they are related. Hence, contrary to Bourdieu's assumption, there is no homology between social positions, and the distances between agents vary accordingly. The status of an agent and the value of his resources may change considerably between different fields and over time.

For the purpose of this study, social as well as specific symbolic resources are of high value, due to the nature of a society at war and the particular history of the field. First of all, as already indicated above, certain resources gain symbolic value for power claims and thereby account for some of the few cases of vertical mobility within the field of resistance. For example, the status of commanders from the group of strongmen was secured and characterized by symbolic resources such as having fought and survived many battles during their lifetime. Although their social background was from a lower peasantry, they managed to become high-ranking commanders due to their combat experience, survival of the Khmer Rouge regime as ordinary Lon Nol soldiers, as well as a spiritual belief in their magical powers, which was regarded as the main source of their strength and thus the reason for their survival. According to Max Weber (1978), this resource could be termed "warrior charisma", which elevates certain social groups in status within violent conflict due to an aura of invulnerability and superior fighting capabilities (Schlichte, 2004, p. 192). A resource like this is difficult to weight within a nation's social structure and, in the end, only makes sense within the field and as long as the field exists.

In addition to symbolic resources, social resources are very important within the field, since they broker important positions as well as access to essential goods while economic means may not be very useful in facilitating a better position within the field. Moreover, within a patrimonial political culture, clientelism determines the bulk of positions within political bodies such as armed groups. Most positions within the field are not contested and largely rest upon a reconstitution of former patrimonial elite networks (within the leadership), their clients (within the mid-range), and a displaced peasantry (among the rank-and-file soldiers). Thereby, one might say, a classic Cambodian political structure was re-established within the guerrilla factions. The point is that patronage networks had not been replaced during wartime but, instead, simply became militarized (cf. Hoffman, 2007, p. 660). Competition over positions only occurred among the leadership groups, and vertical mobility was seen only seldom, such as with the strongmen and some commanders within the mid-range who had gone through the Cadet School. Most positions were allocated within prefixed lines along dyadic loyalties, and people changed positions horizontally: Leaders changed their responsibilities and job descriptions nearly on a yearly basis depending on what they were 'asked to do', while rank-and-file soldiers ameliorated the effects of their bottom position by defection. Therefore, social resources are decisive for the entry into the armed group and for access to basic goods within the camp's economy, but not so much as a 'resource' used to facilitate an improved position in the field, as much of it follows a reproduction of an inherited political status between the refugees.

All four resources matter for the access to as well as mobility within the field and can be used as a resource in competition with others. At the same time, their value fluctuates greatly across time, due to the fact that the possibility of physical harm in war raises the stakes. War may destroy many resources in a very short period of time but still leaves many inequalities between individuals that rested upon former resources that remain untouched. Using a qualitative research design makes it possible to grasp aspects such as major changes and transnationality in lifecourses, the composition and fluctuation in value of resources, and the specific mode of access to the field. Differences are captured that make a difference within the field and within the habitus formation of the respective groups. Furthermore, problems that researchers face in studying a society at war, such as the difficulty of acquiring reliable quantitative data from the field, are met. Even if some data can be collected on the economic and cultural resources circulating within armed groups, the data would not be sufficient for a proper statistical approach (correspondence analysis in particular), although such data may certainly yield some useful information. Moreover, the value of resources may cease or increase quickly, mostly depending on classifications within the field and rules of valuation that may change. Under conditions of war, values change even more dramatically and constantly, making stable valuation almost impossible. Many resources may be of importance for only a very short period of time, such as during the inception of the field. The qualitative methodology tries to do justice to the dynamic and historical foundation of Bourdieu's concepts of habitus, resource, and field, and their application to a transnational field in a society at war. It also has to deal with the fact that the values of resources change according to the field and the oftentimes contested classifications within it. The value of resources is bound to the social relations in which they are embedded, which is also why some of them are highly volatile or almost exclusively relevant within particular fields.

Michel Foucault's theory must also be adjusted significantly. Despite discursive formations, the study also maintains that the Foucaultian power types are socially differentiated. While there can be certain degrees of hegemony for some types, they are nevertheless coexistent and bound to certain habitus groups. While within the Khmer Rouge, disciplinary and pastoral power techniques were highly hegemonic due to the fact that the leadership managed to exclude or simply kill opponents and leave a rather unified habitus group at the top, the non-Communists show highly disparate power techniques at work within their ranks (which may be also one reason for the ongoing leadership quarrels). Each leading habitus group within the KPNLAF and ANS displayed a different habitus and a henceforth different power type structuring its disciplinary practices (which contradicts Foucault's and Bourdieu's rather unified hegemonic elites). In contrast, the Khmer Rouge leadership was highly centralized, without major competition within their ranks. However, even for the Khmer Rouge, there is a social divide that accounts for changes in the only seemingly hegemonic practice and discourse within its ranks, which comes to the fore when studying the movement's mid-range soldiers: Lower ranks reinterpreted the highly intellectualized disciplinary schemes according to a more practical line and thereby reformulated concepts of 'normed practice' into practices of 'learning by doing'. While the leadership developed schemes that were to structure practical experience by prescribed norms that were to be trained in close-knit combat simulations, these concepts oftentimes remained foreign to their subordinates and were implemented, for example, by sending soldiers into battle 'to learn by doing'. Hence, although the leading power practices within the Khmer Rouge were quite hegemonic, their implementation was adjusted by the habitus of the operators. The study maintains that power relations are not just relational forces, as claimed by Foucault, but are social relations that are part of a given society's or a field's social structure.

While the power practices on the part of the leadership groups mirrors specific Foucaultian power types, the practices of the mid-range soldiers do not follow a particular type of power, and the rank-and-file soldiers can be described rather broadly by self-techniques without a specific Foucaultian logic behind their practices. The main reason for this is simply that Foucault always focused on hegemonic discourse formations and power techniques and only included strategies for coping with power at the very end of his writing. Since he did not regard power as socially differentiated, he neither studied the techniques of mid-range power holders nor seriously engaged in the conceptual construction of different types of power 'coming from below and in between'.

In conclusion, the formation of an individual's habitus may be defined by the sum of the agent's social background and/or his family of origin, his upbringing, his acquisition of resources during his lifecourse while participating in different fields and institutions (be it external or as incorporated capabilities), as well as by schemes of perception, thought, and action. The milieu of an individual's family of origin has a decisive impact on the agent's lifecourse and prescribes major directions of his or her lifecourse due to the *hexis* of the habitus, which tries to recreate the conditions of its creation. Later on, events and experiences are already perceived and interpreted by schemes that have been incorporated earlier on and transformed into patterns. War, however, changes life traits and values of resources drastically. In our case, the Cambodian genocide in particular had an important impact on lifecourses of the individuals interviewed and thereby on their habitus and its classificatory schemes. Hence, a field may be defined as 1) a relational space of forces 2) with a certain set of rules (e.g., for accessing the field and mobility within it) and 3) an unequal distribution of power to influence rules and practices, in which 4) agents with varying amounts and compositions of inherited and acquired resources 5) position themselves in relation to others and 6) classify and value others as well as specific resources. There are a couple of important characteristics of fields, with some that are conceptualized slightly differently than within Bourdieu's theory:

1. The valuation of external resources and incorporated capabilities that individuals bring in varies between each field.
   The study puts a stronger emphasis on social and symbolic resources, as in a spiritual "warrior charisma" of the invulnerable, gaining momentum in fields during conflict. Resources can be highly volatile and bound to specific fields.
2. Furthermore, a field is a space of contestation over positions, valuations, and resources.

Contrary to Bourdieu, however, this is not taken as the central characteristic of fields. Although agents position and distinguish themselves in relation to others, this does not necessarily lead to contestation over higher positions. Much to the contrary and amongst others due to the patrimonial structure of the field in question, contestation over positions is rather rarely seen, due to a widespread acceptance of one's 'natural' position (symbolic violence). While for Bourdieu society engages in a constant war all against all, it seems that war is not a state of war.

3. Fields are governed by a central *illusio*, the *nomos*, which in our case would be that expelling the Vietnamese is worth fighting or even – for the lower ranks in particular – dying for.

   Contrary to Bourdieu, agents within the field do not need to full-heartedly believe in the value of this illusio but need to acknowledge it at least formally. Usually, it seems, the lower the individual's status within the field, the more its acknowledgement becomes reduced to mere lip service. Agents may believe that the benefits of being recruited as a soldier outweigh the risks, but maybe simply because in wartime it is safer for them to be a soldier in combat rather than a civilian in between (Nordstrom, 1992). The illusion within the field is always contested and becomes relational and multi-centric as well.

4. Fields are relatively autonomous.

   Due to the transnational habitus formation of the individuals within, the field cannot be subsumed under the Cambodian, the Thai, the French, or the US social structure. Although it stands in relation to all of them and to different other fields, its embeddedness would pose another theoretical and empirical challenge (e.g., an analysis that includes agents from the UN, different states, and ethnic Khmers living in Thai Surin etc.).

5. Fields encompass different habitus groups that share a distinctive discourse and certain types of power that are socially differentiated. Each habitus group, moreover, tries to promote a specific resource, of which they exclusively dispose, to codify their symbolic superiority within the field.

   Incorporated classificatory schemes governing an individual's (power) practices are embedded within the discourse as a symbolic universe. But certain discursive formations and power practices can be more or less hegemonic as well or can be found across various habitus groups (e.g., patronage practices). These discursive elements, however, change their meaning and function between groups, since they are embedded in a different symbolic universe and context of relational positioning. Different power types, moreover, follow a specific logic, some of them resembling what had been described by Foucault (leadership practices in particular). Depending on the respective field in question, different resources can rise or diminish in their symbolic value to legitimize leadership claims. Agents try to define the valuation of the resource they dispose most of (battle experience, education, etc.) to their own benefit.

Conclusion

The study proposes to analyze practices in non-state armed groups as socially rather than spatially (Kalyvas, 2006) or economically (Weinstein, 2007) differentiated. Practices are not just shaped by economic conditions or degrees of control exerted within a certain area but also reflect social structures within fields. They are shaped by the respective classificatory discourse of the agents. The analysis is based upon a slightly adjusted theoretical field approach by Pierre Bourdieu and the conceptualization of power techniques by Michel Foucault. Instead of asking which disciplinary techniques might be effective (van der Haer et al., 2011), it asks why commanders opt for certain practices and not others. And sometimes, it examines why they do *not* opt for practices such as proper trainings, political indoctrination, or combat control. They opt not for what is effective from a general military perspective but for what 'makes sense' to them, what they believe is effective. Most of the current research on civil wars comes from political and economic science, accounting for the bulk of the recent progress in theorizing insurgent and incumbent violence and for a turn towards the micro-politics of armed groups. However, collective violence and war remain at the periphery of sociological thought, and those few exceptions mostly deal with the question of how to integrate violence into the idea of modernity and social order (for an overview, see Malešević, 2010). One reason for this might be that sociological theory seldom examines societies outside of Europe and North America and still largely sticks to the idea that social order is non-violent. Violence is understood as the breakdown of orderly social exchange rather than its continuation. Oftentimes, civil wars are interpreted as the resurfacing of archaic barbarity that undermines modernity (e.g., Kaplan, 2005), not as organized and essentially social processes.

Each of the habitus groups has a different power type that it tries to establish, maintain, or use to undermine rule. The power techniques of each leadership group differ along certain lines, many of which can be described using concepts presented by Michel Foucault. There are intellectuals who use security practices and basic incentives to rule the collective, and effects between different collectivities. There are strongmen who rule by the logic of sovereignty and enforce cohesion by group fraternity. Commanders from the old military elite use military drill as learned in French academies, a mixture between physical modeling and collective humiliation. And finally, there are the anti-intellectual intellectuals who use disciplinary and pastoral techniques to lead their forces. These power types are transmitted by mid-range operators who apply but also adjust these practices according to their own schemes. Most obvious are the changes during transmission by the Khmer Rouge blank-page leaders, who reinterpret normed practice into 'learning by doing'. Pure practice that is not driven by theories serves to instruct the soldiers. At the lower end of the hierarchy, the rank-and-file combatants either support the incorporation of rule by self-techniques, overcoming probable reluctance they may feel regarding orders given to them, or they rely on lines of flight to undermine its worst effects. Lines of flight can be understood quite literally as defection, or rather drastically as executing disagreeable commanders during combat, shooting your own leg to avoid being sent to combat in the first place, simply avoiding an exchange of fire with a friendly enemy, or any other practice that cushions power claims and demands. At most, these techniques merely ameliorate the worst effects of being at the bottom of the hierarchy, without seriously improving the status or living conditions of the rank-and-file soldiers.

All of these techniques are guided not by untimely efficiency or by rational calculus but by a classificatory discourse of the agents. Military leaders opt for such techniques because, for them, these techniques make sense as a tool for ruling their followership while keeping their 'human nature' in mind. Or because they believe that this is the only tool with which they are equipped to deal with the reluctance of soldiers or the unacceptable commands handed down to them. Practices mirror the classificatory scheme that in turn reflects the habitus formation of the agent alongside the volume and composition of the resources of which he disposes. A group's political orientation and ability to recruit soldiers are not sufficient grounds to explain variations within the field, since the differences seem to be similar across different groups and, conversely, may differ sharply between commanders active within the same group. Hence, variations do not solely depend on an agent's membership to a certain formal organization and its political direction. Moreover, degrees of formalization and institutionalization may differ within one group as well. While old military elite commanders strive for formalization of ranks, rules, and military institutions, strongmen undermine any of these attempts to secure the basis of their rule.

Much of the hierarchy within the field reconstitutes the status of agents before joining the resistance – oftentimes either as patrons or their respective clients from the military and political field or those without any social resources at all who end up at the bottom of the insurgency. This also shows the importance of social resources for the structuration of the insurgency and its hierarchies. However, there are two exceptions as a result of the distinct nature of the field. Many agents experienced a lift in status due to the so-called 'warrior charisma' they earned via combat experience or, within the Khmer Rouge in particular, as a result of an inverted valuation of illiteracy and class purity. But even within the Khmer Rouge, the social resources of the top command accelerated the vertical mobility of the so-called blank-page soldiers. When comparing all of the groups within the insurgency, similarities across the field become obvious. For instance, strongmen share a similar family background with the roughnecks and blank-page leaders but differ in terms of education and access to armed groups. These differences in the lifecourses of the individuals make a difference in how they form their habitus but also constitute, as Ludwig Wittgenstein (2009) would term it, a 'family resemblance', in which there is no essential and single common or shared similarity but, rather, a series of overlapping (non-identical) similarities. In general, strongmen completed at least their Bac I, while blank-page leaders were underage monks during recruitment. Roughnecks and blank-page leaders share most of their habitus, which is also why many former Khmer Rouge leaders are among the roughnecks.

The same goes for the old military elite and the loyal tacticians. Here, a strong patron-client relationship forms a connection between the two that is supported by a similar institutional background due to the fact that all of them were trained in military academies of some sort. The difference, however, is that the old military elite comes from known and well-connected military elite families and went to prestigious military academies in Cambodia and abroad. As a result, some notions and schemes are reiterated but are put into a different context and relational position within the field. Moreover, transmission of practices between habitus groups with a similar habitus formation seems to be 'smoother'. While, for instance, tacticians implemented commands and practices in a manner similar to the elite commanders, blank-page leaders adjusted some of their practices a lot when compared to the anti-intellectual intellectuals within their top command. Within the Khmer Rouge, a certain layer of leaders right below the anti-intellectual intellectuals could not be interviewed because they are either dead or in fear of criminal proceedings at the Khmer Rouge Tribunal. Most of these were monks who turned into early rebels for independence with a long record of insurgent action (e.g., Ta Mok, Ke Pauk, etc.). But, as mentioned, while this study was able to uncover these four habitus groups in the field, there could be others.

Leadership groups engrave their respective power types within the field and under their own command in particular. Depending on the respective classificatory scheme and therefore along socially differentiated lines, different Foucaultian power techniques characterize the practices put forward by these groups. While Foucault himself analyzes the rising up and assertion of discursive formations and power types as a relational struggle, he does not frame these relations as social relations in the sense of a society's social structure. Power relations are only explored as relations of power, sometimes leading into a rather circular argumentation, since, as a pure interplay of forces, they are not again bound to anything beyond power: "It seems to me that power must be understood in the first instance as the multiplicity of force relations immanent in the sphere in which they operate and which constitute their own organization" (Foucault, 1991, p. 92). While it may be justifiable in a purely discursive analysis to remain within this framework of fields governed by power forces and figurations of knowledge, this study proposes that they are related to social forces that constitute them and are constituted by these power forces in return. Power is constitutive to a social structure and, at the same time, part of its relations with and formation within different fields.

For Jeremy Weinstein (2007), the rank-and-file soldiers may either be characterized as opportunistic or as active-idealistic. Thereby he opens up a highly simplistic divide between idealists motivated by noble causes that are being pursued by civilized means and investing into the group's well-being and greed-driven opportunists who pursue their own goals by brutish, 'indiscriminate' means and consume the group's benefits. However, beyond the already difficult ethical tone that this argumentation takes, there are some shortcomings to this theory. First, as becomes obvious when interviewing the rank-and-file soldiers of ANS and KPNLAF, both types of combatants may coexist within one group. And second, categories or labels such as greed and grievance are much too simplistic to understand motives, thoughts, and practices of soldiers within the lowest ranks. Since these commanders are bound to satisfy the top command, grievances as master cleavages are constantly recited by the mid-range. During the interviews with the lower ranks, hardly anyone pointed to his will 'to resist'. Those who did clearly spoke full of irony and engaged in laughter afterwards. Although they may have had a negative attitude toward 'the Vietnamese', their motives were much more complex and the trajectory leading them to a refugee encampment much more erratic. At most, joining the 'resistance' came out of necessity – for instance, stemming from the necessity to find a stable source for regular meals due to limitations inherent in the UN's distribution system. The attitude of the pragmatists, moreover, was not greed but, rather, the result of their adaption to the realm of and life within an armed group.

In order to respond to realities in the field, Pierre Bourdieu's approach had to be adjusted at certain points. The structure of and hierarchy within the insurgency has to include more than cultural and economic resources, with economic resources even diminishing or becoming highly ambiguous in practice within a society at war. In the Cambodian case at least, social resources gained momentum during wartime, with a displaced peasantry looking for patrons who could care of them after the rupture of kinship ties and the militarization of older patrimonial networks in politics. In addition to a structuration by patrimonial reproduction of a trinity of patrons, clients, and a displaced peasantry, symbolic resources defined by the rules of valuation within the field become decisive as well. Warrior charisma, in particular, which a soldier earned by surviving the genocide and several battles, gains momentum within the movement and elevates many agents to rather high positions, after having lived lives that were characterized by comparatively low social status gains, if any at all. But these gains always remained threatened by their volatility and by possible peace talks. Moreover, the lifecourse of any soldiers needs to be taken into account when examining his habitus formation as well. The lifecourse does not simply fulfill the prescribed social fate of the agents; the point in time when agents enter the field and when they earn which kind of resources are decisive. They do not simply become what they are after birth or as a result of the social position they inherited within their families. This becomes clear, for instance, when agents who have a similar social background and occupy a similar position in the insurgency still do not have the same habitus – like in the rank and file or when strongmen earn decisive symbolic resources after years of being ordinary rank-and-file soldiers.

The rank-and-file soldier is a good example of the difference that events in one's lifecourse can make. The main differentiating factor for those being interviewed for this project is their age of recruitment or of becoming a belligerent and displaced by war. This shows that participation in different fields during one's lifecourse and the respective mode and timing of this participation have a decisive impact on habitus formation. Thus, these factors need to be considered in the analysis. At least in this case, a simple divide between greed-driven opportunists and grievance-driven activists, as is used in some older and many recent studies, does not suffice to describe the recruits' motives and behavioral schemes. Yet another example are the anti-intellectual intellectuals within the Khmer Rouge leadership, who developed large parts of their 'heretic discourse' as up-and-coming forces within the political field, when coming under the sponsorship of the intellectual circles within the Democratic Party during the 1950s and 1960s. These years were decisive in constituting a heretic classification and idea of society, rule, and politics that ultimately structured the usage of power practices and institutions within the NADK as military force. Moreover, due to the double function of the Central Committee as political and military leadership, this certainly applied to other domains of their rule as well. To understand the power techniques and the (re-)constitution of rule within the insurgency, therefore, means to examine the habitus formation of agents while taking their inherited social background into account as well as the resources they acquired during their lifecourse and during the participation in different fields. This is what a sociological approach to non-state armed groups may illuminate.

Given the change in warfare, coined as an age of irregular warfare within so-called 'new wars', a deeper analysis of the social composition of armed groups might point towards a better understanding of their behavior. Since many groups are non-state entities who, like in our case, comprise only few military professionals, their behavior becomes increasingly uncoupled from state military socialization and regulation. These groups may encompass fighting intellectuals, politicians, farmers, workers, and any other agent one might imagine. Thus, demobilization is more than a peace agreement between top commanders and a reintegration into a state military, although this might be the case for many. However, demobilization and reintegration need to address different the social positions and lifecourses of belligerents. A warlord, for instance, may 'spoil' peace negotiations simply because he has too much to lose after gaining many of the resources that define his status in wartime. For these agents, transformation of resources into resources that are still relevant during peace will certainly be their main interest. Already within the field itself, the resources of the strongmen are especially volatile, which is why they have an eye on transforming them into more stable and sustainable resources. Many of the strongmen interviewed for this project tried to secure (mostly successfully) a position within ministries or as provincial governors. Moreover, many had to make sure that they would not face judicial proceedings after the end of the war. While strongmen certainly are most likely to be the ones who spoil peace talks and proceedings due to the volatility of their position, others – like the intellectuals – quickly reintegrated, since they had a lot of legitimate resources at hand that would be relevant during times of peace; in the end, most of them could even enter slightly higher political offices. Instead of labeling some agents as co-operative and others as 'brutish' or 'greedy', one should understand the background of and reasons for the choices that belligerent soldiers made during demobilization and reintegration – choices that each soldier and commander within the field made during its collapse.

The study yielded several theoretical and empirical results that are important for a sociology of civil wars and conflict and that could be summarized as follows:

- Power practices are routed in the habitus of agents, and not in economic or spatial conditions.
- The field of insurgency reproduces social differentiations prior to its own formation. Many positions in the field are homologous with older hierarchies, e.g., with an old military and political elite forming its upper ranks, their patrimonial clients within the mid-range leadership, and a displaced peasantry within the rank and file. This shows that social differentiations survive even massive societal changes, with political networks being reconstituted and militarized.
- Each habitus group needs to adapt to the new situation by structuring its practices through a classificatory discourse on leadership and soldierhood. Agents are differently well prepared to handle this transition.
- However, a field cannot be explained solely by reference to a prior social structure being reproduced; its own historical and symbolical formation are also essential. Each field has a different history and symbolic universe, which values resources differently. Therefore, some groups are able to rise in status. In our case, this would be the valuation of combat experience within an insurgency. This enables groups such as the guerrilla strongmen, the battle-hardened roughnecks, or the Khmer Rouge's blank-page leaders to rise in status.
- These resources, however, are highly field-bound and volatile. This affects not only the willingness of agents to end the war due to the inherent risk of falling back into a lower status in a society at peace but also their power practices within the field. Groups whose claim to power rests upon highly volatile symbolic resources need to guard and reconstitute their sovereignty at every step. That is why their power practices resemble what Foucault called 'sovereign practices'. Others with more stable resources, such as the intellectuals, do not make much of an effort to secure their leadership position and skip large parts of the disciplinary power apparatus aimed at making soldiers responsive to orders. They even tend to ignore when soldiers refuse their orders.
- Last but not least, the habitus of agents proves to be lethargic. This means that their schemes of thought, perceptions, and actions are changed not at random but according to a rational calculus on the part of homo economicuses. While wars can be characterized as massive societal changes carrying dynamics of their own, which affect power relations and social structures, the agents' practices during the transition and within the new situation are guided by a longue durée of habitus formation, in which basic patterns of behavior have been incorporated into and are evolving in the individuals' bodies. The only

notable exception is child soldiers, simply due to the fact that they have been socialized within the field and thereby incorporated basic behavioral patterns of the armed group to which they belonged.

Post-Script: A Note on the Dissolution of the Field

While the upper echelon of the guerrilla force was either integrated by Premier Hun Sen's *win win policy*, securing among others high-ranking government positions, defeated in a coup in 1997, or vanished from the political arena during the *United Nations Transitional Authority in Cambodia* (UNTAC) and its subsequent elections (1992–1993), the transition to peace was experienced very differently for those below. During interviews, it was striking to see where the former belligerents made their living today. Because some of them followed patrons or because they controlled a considerable number of soldiers, some became high-ranking government officials or commanders in the *Royal Cambodian Armed Forces* (RCAF). Many from the non-Communist leadership found support as leaders of NGOs financed by foreign governments and are important figures in today's Cambodian civil society. Others, however, are impoverished nowadays and fed up with what happened to them after the war and reintegration, and they went on with resistance as long as possible – especially the roughnecks from the mid-range, of whom many could not transfer their symbolic resource as tough warriors into relevant resources in peacetime. Similar to the pragmatists, they struggled with building up a life after war, simply due to the fact that they hardly ever experienced life in times of peace. Additionally, the government did not keep many promises, nor did the subsequent political parties of the military factions or the UN organizing repatriations. A large number of refugees from the camps were relocated to two villages – one close to Battambang and one close to Pursat. These villages are isolated, and people living there face high poverty as well as stigmatization by Cambodian society for being those who fled the country. Proper research on post-war transition and biographical trajectories of belligerents is still pending.

All that is known so far is that the UN staged two attempts to reintegrate former combatants. The first was during the UNTAC period. Here the plan was to demobilize 70 per cent of 200,000 soldiers from all four factions (including the government). Yet, due to the withdrawal of the Khmer Rouge from the peace process, UNTAC failed to disarm the forces. They were badly needed to keep the Khmer Rouge at bay, since occupation troops from other countries were not willing to engage in civil war instead of peacekeeping operations. As a result, the troops of the Cambodian state were reduced only by 9,003, ANS by 3,187, and KPNLAF by 1,322 soldiers (CICS, 2008, p. 9). The idea was to give each combatant an identity card, a health screening, initial counseling, a 'safety package' of cash (240 US dollars), food, materials to build a small wooden house, and transportation to their home communities. In the end, however, donors considered it impossible to distribute on such a large scale and therefore distributed 253 sets for building a shelter, 872 sets of generators/wire/water pumps, 111 sets of generators/wire/

sewing machines, and 13,764 motorbikes/spares/sewing machines. All received a medical kit and some skills/vocational training and job counseling as well (CICS, 2008, p. 12). The World Bank considered the outcome unsatisfactory and sustainability unlikely. Furthermore, distribution reached only half of the combatants during the first year.

Many belligerents, however, blame themselves for shortsightedness during repatriation at a time in which they had the choice to take either money to start a business or land for farming:

> I am currently in a bad living condition as I am renting a house [a wooden shack poorly cobbled together, which he is currently renting for five dollars a month]. I decided to take money, not land after repatriation. UNHCR was giving us land back then, but I was so short-term oriented that I decided to take the money in an attempt to open a business. However, soon the entire money was used up. (GC-KP1)

As a former child soldier and pragmatist experiencing barely more than war, he was too shortsighted and opted for an immediate and monetary benefit. While it seems that former child soldiers largely opted for the money, spiritualists returned 'home' to do farming. The question for demobilization programs, then ,was whether it would be better in the long run to provide vocational training rather than monetary incentives, which may be useful for facilitating a quick disarmament but may hamper proper and sustainable reintegration. There are few studies on the factors influencing decisions of former belligerents during post-conflict transitions. Most of these studies, moreover, focused on leadership transitions and not on what happened to the whole field of resistance (e.g., Gerdes & Hensell, 2012). The question remains why some were able to transfer their own field position into a similar position within a new field after war, or even improve their status and others faced a stark downturn in status. What determines the trajectories of lives after the end of war? What shapes belligerents' choices during reintegration? It seems likely that agents with highly field-specific symbolic resources are the ones who will 'spoil' peace negotiations. Labeling them as 'greedy' does not explain the logic behind peace transitions (e.g., Stedman, 2008). Theorizing the logic behind power practices within armed groups might be crucial to devising an approach to demilitarize these entities and, thereby, foster conditions for long-term peace as well.

Appendix I:
Map of UNHCR-UNBRO Border Camps during the 1980s and 1990s

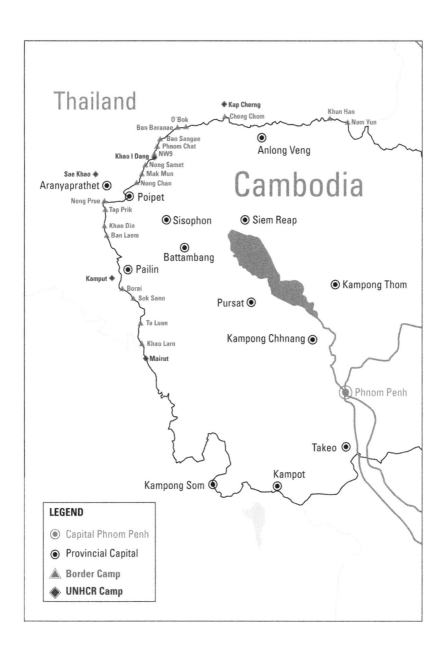

Thailand

Cambodia

♦ Kap Cherng
O`Bok
Chong Chom
Khun Han
Nam Yun
Ban Baranae
Ban Sangae
Phnom Chat
Khao I Dang NW9
Nong Samet
Anlong Veng
Sae Khao ♦
Mak Mun
Aranyaprathet ⊙
Nong Chan
Nong Prue
Poipet
Tap Prik
Khao Din
Sisophon
Siem Reap
Ban Laem
Battambang
Pailin
Kamput ♦
Borai
Sok Sann
Kampong Thom
Pursat ⊙
Ta Luan
Khao Larn
Kampong Chhnang ⊙
♦ Mairut
Phnom Penh
Takeo ⊙
Kampot
Kampong Som ⊙

LEGEND
⊙ Capital Phnom Penh
⊙ Provincial Capital
▲ Border Camp
♦ UNHCR Camp

Appendix II:
Questionnaire

The questionnaires for habitus hermeneutics are divided into three parts. Each of the three semi-structured parts comprises two or three open question, and some further inquiries, which could be used optionally depending on the course of the interview. Questions on habitus formation and lifecourse are largely the same for leadership, mid-range, and lower rank, but differ when talking about the classificatory discourse and power practices. The open questions are designed to suffice for covering all topics, at least ideally.

Part I: Social Background, Acquired Resources and Lifecourse

Open questions:

- First, we would like to know more about your life. Could you, please, start by telling us a bit about your family of origin?
- Now we would like to know more about how you became a member of the resistance. Please describe what happened.

Optional inquiries:

- Family background and childhood
- Schooling
- Khmer Rouge years
- Experience during Vietnamese invasion
- Flight to the border
- Social resources within the field
- Reason for joining the particular faction/commander
- Access limitations

Part II: Classificatory Discourse

Leadership

Open questions:

- Could you please describe your work as a commander in your unit?
- How would you describe a typical rank-and-file soldier of the resistance?

Optional inquiries:

- Concept of a successful commander
- Demands of recruits for their loyalty
- Characteristics of their own leadership style
- Characteristics of rank-and-file soldiers
- Trainings for whom and why?
- Recruits changed by service?
- The military academy and its structure

Mid-range operators

Open questions:

- Could you please describe your work as a commander in your unit?
- Could you please describe everything that a soldier needs to be ready for combat in your opinion?

Optional inquiries:

- Concept of a successful commander
- Demands of recruits for their loyalty
- Characteristics of their own leadership style
- Becoming skillful
- Preparations for combat
- Military, moral, and political training
- Instructions by monks

Rank-and-file soldiers

Open questions:

- How would you describe your commander?
- Could you please describe everything that a soldier needs to be ready for combat in your opinion?

Optional inquiries:

- Further classification of commander
- Characteristics of good leadership
- Comparison of different commanders
- Modes of and criteria for defection
- Becoming skillful
- Preparations for combat
- Use of military, moral, and political trainings

- Instructions by monks

Part III: Analysis of Power Practices

Leadership

Open questions:

- Could you please describe what fresh rank-and-file recruits need to be ready for combat?
- Could you please describe how it could be ensured that soldiers actually followed the orders given for and also during combat operations?
- Could you please explain how indiscipline was dealt with?

Optional inquiries:

- Preparation for combat
- Military, moral, and political trainings
- Successful and good soldiers
- Ensuring that a group actually went to battle
- Ensuring that individual soldiers actually fought in battle
- Overcoming reluctance of soldiers
- Post-combat rectification and reporting
- Indiscipline according to the ranks
- Effectiveness of certain rewards and punishments
- Source of indiscipline

Mid-range operators

Open questions:

- Could you please describe what the easiest way to earn respect as a soldier was?
- Could you describe how it could be ensured that soldiers actually followed orders given for and also during combat operations?
- Could you please explain how indiscipline was dealt with?

Optional inquiries:

- Respect within unit
- Respect of superiors
- Behavior when facing disrespect
- Self-classification
- Good life as a soldier
- Ensuring that a group actually went to battle

- Ensuring that individual soldiers actually fought in battle
- Overcoming reluctance of soldiers
- Post-combat rectification and reporting
- Indiscipline according to the ranks
- Effectiveness of certain rewards and punishments
- Source of indiscipline

Rank-and-file soldiers

Open questions:

- Could you please describe what the easiest way to earn respect as a soldier was?
- Have you always been committed to fighting? On which occasions were you not and why? Please describe what happened.

Optional inquiries:

- Respect within unit
- Respect of superiors
- Behavior when facing disrespect
- Self-classification
- Good life as a soldier
- Unwillingness due to fear
- Others having fear, reasons
- How to work on or get over your fear
- Unwillingness due to moral dilemma
- Avoidance behavior
- Unwillingness due to illegitimacy of command

Biographical Checklist

At the end of each interview, further key biographical data are reviewed in case they are not already clearly mentioned during the interview. This is done to ensure that no essential information is missing:

- Current age, place of birth, highest educational level reached so far.
- Which unit, battalion, brigade?
- How many people did you command?
- Who was the commander of your unit, brigade (top leader)?
- Where were you stationed? Which camp or area?
- Rank when joining the military? Rank when leaving? Which ranks in between and for how long?

Bibliography

Ayres, D. (2000) Tradition, Modernity, and the Development of Education in Cambodia. *Comparative Education Review*. 44 (4). pp. 440–63.

Azar, E. (1990) *The Management of Protracted Social Conflict: Theory and Cases*. Hampshire: Dartmouth.

——— (1991) The Analysis and Management of Protracted Social Conflict. In: Volkan, V. & Montville, J. (eds). *The Psychodynamics of International Relationships*. Lexington: D.C. Heath.

Ballentine, K. & Sherman, J. (eds) (2003) *The Political Economy of Armed Conflict: Beyond Greed and Grievance*. Boulder: Lynne Rienner.

Becker, E. (1998) *When the War was Over*. New York: Public Affairs Press.

Bekaert, J. (1991) Playing the Numbers Game with Cambodia's Paper Armies. *Jane's Defence Weekly*. November 2, 1991. p. 841.

——— (1997) *Cambodian Diary. Tales of a Divided Nation, 1983–1986*. Bangkok: White Lotus Press.

——— (1998) *Cambodian Diary. A Long Road to Peace, 1987–1993*. Bangkok: White Lotus Press.

Berdal, M. & Melone, D. (2000) *Greed and Grievance: Economic Agendas in Civil War*. Boulder: Lynne Rienner.

Boua, C. et al. (eds.) (1988) *Pol Pot Plans the Future: Confidential Leadership Documents from Democratic Kampuchea, 1976–1977*. New Haven: Yale University Press.

Bourdieu, P. (1977) *Outline of a Theory of Practice*. Cambridge: Cambridge University Press.

——— (1986) The Forms of Capital. In: Richardson, J. (ed.) *Handbook of Theory and Research for the Sociology of Education*. New York: Greenwood. pp. 241–58.

——— (1990) *The Logic of Practice*. Stanford: Stanford University Press.

——— (1993) The Field of Cultural Production. In: Johnson, R. (ed.) *The Field of Cultural Production: Essays on Art and Literature*. New York: Columbia University Press.

——— (1996) Vilhelm Aubert lecture: Physical Space, Social Space and Habitus. Oslo: University of Oslo & Institute for Social Research.

——— (1997) Für einen anderen Begriff von Ökonomie. In: Bourdieu P. (ed.) *Der Tote packt den Lebenden*. Hamburg: VSA. pp. 79–100.

——— (2010) *Distinction: A Social Critique of the Judgement of Taste*. London/ New York: Routledge.

——— & Wacquant, L. (1992) *An Invitation to Reflexive Sociology*. Cambridge: Polity Press.

Bremer, H. (2004) *Von der Gruppendiskussion zur Gruppenwerkstatt*. Münster/ Hamburg/London: Lit.

Bultmann, D. (2012) Irrigating a Socialist Utopia: Disciplinary Space and Population Control under the Khmer Rouge, 1975–1979. *Transcience*. 3 (1). pp. 40–52.

Cambodian Genocide Program (CGP) (2010) Autobiography of Thiounn Prasith. Dated on December 25, 1976. Available at <http://www-yale.edu/cgp/thiounn. html>.

Carney, T. (ed.) (1977) *Communist Party Power in Kampuchea (Cambodia): Documents and Discussion*. Ithaca, New York: Cornell University Press.

Center for International Cooperation and Security (CICS) (2008) Disarmament, Demobilisation and Reintegration (DDR) and Human Security in Cambodia. University of Bradford.

Central Intelligence Agency (CIA) (1987) The 'New Face' of the Khmer Rouge: Implications for the Cambodian Resistance. Declassified document. Available at <www.foia.cia.gov>.

———— (1989) The Non-Communist Factions in Cambodia: The Challenges Ahead. Declassified document. Available at <www.foia.cia.gov>.

Chanda, N. (1986) *Brother Enemy: The War after the War*. San Diego: Harcourt.

Chandler, D. (1983) Revising the Past in Democratic Kampuchea: When Was the Birthday of the Party? Notes and Comments. *Pacific Affairs*. 56 (2). pp. 288–300.

———— (2000) *Brother Number One: A Political Biography*. Chiang Mai: Silkworm Books.

———— (2008) *The Tragedy of Cambodian History*. New Haven: Yale University Press.

Collier, P. (2000) Doing Well out of War: An Economic Perspective. In: Berdal, M. & Melone, D. (eds). pp. 91–111.

———— (2007) *The Bottom Billion: Why The Poorest Countries Are Failing And What Can Be Done*. Oxford/New York: Oxford University Press.

———— & Hoeffler, A. (2001) Greed and Grievance in Civil War. *World Bank Working Paper*. No. 28126.

Conboy, K. (2013) *The Cambodian Wars: Clashing Armies and CIA Covert Operations*. Lawrence: University Press of Kansas.

Corfield, J. (1991) A History of the Cambodian Non-Communist Resistance 1975–1983. *Working Paper*. Monash University, Australia.

Deleuze, G. & Guattari, F. (2004) *A Thousand Plateaus: Capitalism and Schizophrenia*. London/New York: Continuum.

Demmers, J. (2012): Theories of Violent Conflict: An Introduction, London/New York: Routledge.

Deth, S. (2009) The Geopolitics of Cambodia During the Cold War Period. *Explorations*. 9. pp. 47–53.

———— (2009a) *The People's Republic of Kampuchea 1979–1989: A Draconian Savior?* M.A. Thesis. Athens, Ohio: Ohio University.

Duffield, M. (2002) Social Reconstruction and the Radicalization of Development: Aid as a Relation of Global Liberal Governance. *Development and Change*. 33 (5). pp. 1049–71.

Ea, M. (2005) *The Chain of Terror: The Khmer Rouge Southwest Zone Security System*. Phnom Penh: DC-Cam.

———— & Sim, Sorya (2001) *Victims and Perpetrators? Testimonies of Young Khmer Rouge Comrades*. Phnom Penh: DC-Cam.

Ebihara, M. (1990) Revolution and Reformulation in Kampuchean Village Culture. In: David A. & Hood, M. (eds) *The Cambodian Agony*. New York: Sharpe. pp. 16–61.

Elwert, G. (1999) Markets of Violence. In: Elwert et al. (eds) *Dynamics of Violence*. Berlin: Duncker & Humblot.

Etcheson, C. (1984) *The Rise and Demise of Democratic Kampuchea*. Boulder: Westview Press.

———— (1987) Civil War and the Coalition Government of Democratic Kampuchea. *Third World Quarterly*. 9 (1). pp. 187–202.

Evans, G. & Rowley, K. (1990) *Red Brotherhood at War: Vietnam, Cambodia and Laos since 1975*. New York/London: Verso.

Extraordinary Chambers in the Courts of Cambodia (ECCC) (2010) *Closing Order Case002*. Phnom Penh. Available at <www.eccc.gov.kh/en/documents/court/closing-order>.

———— (2012) *Transcript of Trial Proceedings: Case 002*. Trial Day 58. April 30, 2012.

Falge, C. (2011) *The Global Nuer: Transnational Livelihoods, Religious Movements and War: Integration and Conflict Studies*. Halle, Saale: Max Planck Institute for Social Anthropology.

Fearon, J. & Laitin, D. (2000) Violence and the Social Construction of Ethnic Identity. *International Organization*. 54 (4). pp. 845–77.

Foucault, M. (1970) *The Order of Things: An Archeology of the Human Sciences*. New York: Vintage Books.

———— (1981) The Order of Discourse. In: Young, R. (ed.) *Untying the Text*. Boston/London/Henley: Routledge & Kegan Paul.

———— (1981) 'Omnes et Singulatim': Towards a Criticism of 'Political Reason'. In: McMurrin, S. (ed.) *The Tanner Lectures on Human Values, II*. Salt Lake City: University of Utah Press. pp. 225–54.

———— (1990) *The History of Sexuality: Volume 1*. New York: Vintage.

———— (1991) *Discipline and Punish: The Birth of the Prison*. London: Penguin Books.

———— (1992) *The Use of Pleasure: The History of Sexuality: Volume 2*. London: Penguin Books.

———— (2007) The Meshes of Power. In: Crampton, J. & Elden, S. (eds) *Space, Knowledge and Power*. Burlington/London: Ashgate. pp. 153–62.

———— (2009) *Security, Territory, Population. Lectures at the Collège de France, 1977–78*. New York: Palgrave Macmillan.

French, L. (1994) *Enduring Holocaust, Surviving History: Displaced Cambodians on the Thai-Cambodian border, 1989–1991*. PhD thesis. Cambridge, Massachusetts: Harvard University.

Gerdes, F. & Hensell, S. (2012) Elites and International Actors in Post-war Societies: The Limits of Intervention. *International Peacekeeping*. 19 (12). pp. 154–69.

Giuchaoua, Y. (2009) Who Joins Ethnic Militia? A Survey of the Oodua People's Congress in Southwestern Nigeria. *Crise*. Working Paper. 44.

Goffman, E. (1974) *Frame Analysis: An Essay of the Organization of Experience*. New York: Harper Colophon.

Goodhand, J. (2003) Enduring Disorder and Persistent Poverty: A Review of the Linkages between War and Chronic Poverty. *World Devlopment*. 31 (3). pp. 629–46.

Greve, H. (1987) Evacuation Sites for Kampucheans in Thailand: Virtual Concentration Camps under International Auspices? *Refugee Studies Centre*. November 17, 1987.

Guerrilla Training Manual (2003) In: *Searching for the Truth*. Third Quarter 2003. pp. 16–18.

Gurr, T. (2011) *Why Men Rebel*. Boulder: Paradigm Publishers.

Harris, I. (2005) *Cambodian Buddhism. History and Practice*. Honolulu: University of Hawai'i Press.

Heder, S. (1980) Kampuchean Occupation and Resistance. *Asian Monographs*. 27. Institute of Asian Studies. Bangkok: Chulalongkorn University.

——— (1996) The Resumption of Armed Struggle by the Party of Democratic Kampuchea. In: Heder, S. & Ledgerwood, J. (eds) *Propaganda, Politics, and Violence in Cambodia*. New York: M.E. Sharpe. pp. 114–33.

——— (2012) Communist Party of Kampuchea Policies on Class and on Dealing with Enemies Among the People and Within the Revolutionary Ranks, 1960–1979: Centre, Districts and Grassroots. *School of Oriental and African Studies*. April 26, 2012.

Hinton, A.L. (1998) A Head for an Eye: Revenge in the Cambodian Genocide. *American Ethnologist*. 25 (3). pp. 352–77.

——— (2004) *Why Did They Kill? Cambodia in the Shadow of Genocide*. Berkeley: University of California Press.

Hoffman, D. (2007) The Meaning of a Militia: Understanding the Civil Defence Forces of Sierra Leone. *African Affairs*. 106 (425). pp. 639–62.

Holsti, K. (1996) *The State, War, and the State of War*. Cambridge: Cambridge University Press.

Huy, V. (2001) Khmer Rouge Revolutionary Youth Academy. *Searching for the Truth*. DC-Cam. Number 18. pp. 19–21.

——— (2003) *The Khmer Rouge Division 703*. Phnom Penh: DC-Cam.

Ignatieff, M. (1999) *The Warrior's Honor*. London: Vintage.

International Committee of the Red Cross (ICRC) (1999) Country Report: Cambodia

——— ICRC Worldwide Consultation on the Rules of War. Report by Greenberg Research. Geneva.

Jean, F. & Rufin, J. (eds) (1999) *Ökonomie der Bürgerkriege*. Hamburg: Hamburger Edition.

Kaldor, M. (1999) *New and Old Wars: Organized Violence in a Global Era*. Cambridge: Polity Press.

Kalyvas, S. (2001) Research Note: 'New' and 'Old' Civil Wars: A Valid Distinction. *World Politics*. 54. pp. 99–118.

———— (2003) The Ontology of 'Political Violence': Action and Identity in Civil Wars. *Perspectives on Politics*. 1 (3). pp. 375–494.

———— (2004) Ethnicity and Civil War Violence: Micro-level Empirical Findings and Macro-level Hypotheses. Unpublished.

———— (2006) *The Logic of Violence in Civil War*. Cambridge: Cambridge University Press.

———— & Kocher, M. (2007a) Ethnic Cleavages and Irregular War: Iraq and Vietnam, Politics and Society. *Politics & Society*. 35 (2). pp. 183–223.

———— & Kocher, M. (2007b) How Free is 'Free-Riding' in Civil Wars? Violence, Insurgency, and the Collective Action Problem. *World Politics*. 59 (2). pp. 177–216.

Kaplan, R. (2001) *The Coming Anarchy: Shattering the Dreams of the Post Cold War*. London: Vintage.

———— (2005) *Balkan Ghosts: A Journey Through History*. New York: Picador.

Karniol, R. (1990), Revival of the Khmer Rouge. *Jane's Defence Weekly*. January 20, 1990. pp. 107–8.

Keen, D. (1998) *The Economic Function of Violence on Civil Wars*. Adelphi Paper 320. London: International Institute of Strategic Studies.

———— (2008) *Complex Emergencies*. Cambridge: Polity Press.

Khmer Rouge file (1982) [no title]. Captured in Pailin by government troops. Translated document provided by Kem Sos.

Kiernan, B. (1996) *How Pol Pot Came to Power*. New Haven: Yale University Press.

———— (2002) *The Pol Pot Regime: Race, Power, and Genocide in Cambodia under the Khmer Rouge, 1975–79*. New Haven/London: Yale University Press.

———— & Owen, T. (2006) Bombs Over Cambodia. *The Walrus*. October 2006. pp. 62–9.

King, C. (2004) The Micropolitics of Social Violence. *World Politics*. 35. pp. 431–55.

Kong, T. (2009) Khmer People's National Liberation Front and the Road to Peace. Unpublished manuscript provided by the author. Phnom Penh.

KPNLAF (1987a) Armed Political-Psychological-Clandestine Operations: Internal document from the General Staff. July 22, 1987. Provided by Kong Thann.

———— (1987b) Operation SSD: Internal Document from the General Staff. July 27, 1987. Provided by Kong Thann.

———— (1987c) The "PLANA" Department of the KPNLAF General Staff: Internal document. August 23, 1987. Provided by Kong Thann.

Lange-Vester, A. (2007) *Habitus der Volksklassen*. Münster: LIT.

Lawyers Committee (1987) *Seeking Shelter: Cambodians in Thailand: A Report on Human Rights*. New York: Lawyers Committee for Human Rights.

———— (1990) *Kampuchea: After the Worst*. New York: Lawyers Committee for Human Rights.

Locard, H. (2000) The Khmer Rouge Gulag: 17 April 1975 – 7 January 1979. Phnom Penh. Unpublished.

――― (2004) *Pol Pot's Little Red Book: The Sayings of Angkar*. Chiang Mai: Silkworm Books.

Ly, S. (2002) Social Classes in Democratic Kampuchea. *Searching for the Truth*. 34. pp. 14–17.

Malešević, S. (2010) *The Sociology of War and Violence*. Cambridge: Cambridge University Press.

Maloy, J. (2010) With art as their armor. *globalpost.com*. Available at <http://www.globalpost.com/dispatch/asia/090902/ancient-khmer-tattoos-powerful-dying-art>.

Mason, L. & Brown, R. (1983) *Rice, Rivalry and Politics: Managing Cambodian Relief*. London: University of Notre Dame Press.

Münkler, H. (2005) *The New Wars*. Cambridge: Polity Press.

Murashima, E. (2009) The Young Nuon Chea in Bangkok (1942–1950) and the Communist Party of Thailand. *Journal of Asia-Pacific Studies*. 12. pp. 1–42.

Nhek, B. (1998) A Single Piece of Luck among a Thousand Dangers. Phnom Penh. Unpublished Auto-biography.

Nordstrom, C. (1992) *The Paths to Domination, Resistance, and Terror*. Berkeley: University of California Press.

――― (2007) *Global Outlaws: Crime, Money and Power in the Contemporary World*. Berkeley: University of California Press.

Notebook of Comrade Iv Bun Chhoeun (2001) Temporary Statutes of the Communist Youth League of Kampuchea of the Communist Party of Kampuchea. *Searching for the Truth*. 13. pp. 10–14.

Peang-Meth, A.G. (1990) A Study of the Khmer People's National Liberation Front and the Coalition Government of Democratic Kampuchea. *Contemporary Southeast Asia*. 12. pp. 172–85.

Peschoux, C. (1992) The 'New' Khmer Rouge: Reconstruction of the Movement and Reconquest of the Villages (1979–1990). Paris. Unpublished English translation.

Pol Pot (1988) Report of Activities of the Party Center According to the General Political Tasks of 1976 [20 December 1976]. In: Boua, C. et al. (eds). pp. 182–216.

Picq, L. (1989) *Beyond the Horizon: Five Years with the Khmer Rouge*. New York: St. Martins Press.

Procknow, G. (2011) *Recruiting and Training: Genocidal Soldiers*. Regina: Francis & Bernard.

Rehbein, B. (2003) 'Sozialer Raum' und Felder. Mit Bourdieu in Laos. In: Rehbein et al. (eds) *Pierre Bourdieus Theorie des Sozialen*. Konstanz: UVK. pp. 77–96.

――― (2006) *Die Soziologie Pierre Bourdieus*. Konstanz: UVK.

――― (2007) *Globalization, Culture and Society in Laos*. New York: Routledge.

――― (2008) The Analysis of Global Social Structure. Online publication. Available at <http://www.iaaw.hu-berlin.de/transformation/forschung/material/habitus-hermeneutics>.

――― (2011) Differentiation of Sociocultures, Classification, and the Good Life in Laos. *Journal of Social Issues in Southeast Asia*. 26 (2). pp. 277–303.

Reno, W. (1998) *Warlord Politics and African State*. Boulder: Lynne Rienner.

Reynell, J. (1989) *Political Pawns: Refugees on the Thai-Kampuchean Border*. Oxford: Refugee Studies Programme.

Robinson, C. (2000) Refugee Warriors at the Thai-Cambodian Border. *Refugee Survey Quarterly*. 19 (1). pp. 23–37.

Rogge, J. (1990) *Return to Cambodia: The Significance and Implications of Past, Present and Future Spontaneous Repatriations*. Dallas: The Intertect Institute.

Rowley, K. (2007) Second Life, Second Death: The Khmer Rouge after 1979. In: Cook, S. (ed.) *Genocide in Cambodia and Rwanda: New Perspectives*. New Brunswick, New Jersey: Transaction Publishers. pp. 191–213.

Sanín, F. (2008) Telling the Difference: Guerrillas and Paramilitaries in the Colombian War. *Politics & Society*. 36 (1). pp. 3–34.

Schier, P. (1985) *Prince Sihanouk on Cambodia*. Hamburg: Institut für Asienkunde.

Schlichte, K. (2003) Profiteure und Verlierer von Bürgerkriegen: Die soziale Ökonomie der Gewalt. In: Ruf, W. (ed.) *Politische Ökonomie der Gewalt: Staatszerfall und die Privatisierung von Gewalt und Krieg*. Opladen: VS. pp. 124–43.

——— (2004) Krieg und bewaffneter Konflikt als sozialer Raum. In: Kurtenbach, S. & Lock, P. (eds) *Kriege als Überlebenswelten*. Bonn: Dietz. pp. 184–99.

——— (2009) *Shadow of Violence: The Politics of Armed Groups*. Frankfurt: Campus.

——— (2012) The Limits of Armed Contestation: Power and Domination in Armed Groups. *Geoforum*. 43. pp. 716–24.

Scott, J. (1976) *The Moral Economy of the Peasant*. New Haven: Yale University Press.

Shepler, S. (2005) Globalizing Child Soldiers in Sierra Leone. In: Maira, S. & Soep, E. (eds) *Youthscapes*. Philadelphia: University of Pennsylvania Press. pp. 119–33.

Short, P. (2004) *Pol Pot: The History of a Nightmare*. London: John Murray.

Showcross, W. (1984) *The Quality of Mercy: Cambodia, Holocaust and Modern Conscience*. New York: Simon and Schuster.

Snow, D. & Benford, R. (1992) Master Frames and Cycles of Protest. In: Morris, A. & Mueller C. (eds) *Frontiers in Social Movement Theory*. New Haven: Yale University Press. pp. 133–55.

Stedman, S. (2008) Peace Processes and the Challenges of Violence. In: Darby, J. & Mac Ginty, R. (eds) *Contemporary Peacemaking: Conflict, Peace Processes and Post-War Reconstruction*. London: Palgrave MacMillan. pp. 147–58.

Stuart-Fox, M. & Ung, B. (1998) *The Murderous Revolution: Life and Death in Pol Pot's Kampuchea*. Bangkok: Orchid Press.

Tarrow, S. (1998) *Power in Movement, Social Movements and Contentious Politics*. Cambridge: Cambridge University Press.

Unger, D. (2003) Humanitarian Assistance in Cambodia. In: Stedman, S. & Tanner, F. (eds) *Refugee Manipulation: War, Politics, and the Abuse of Human Suffering*. Washington: Brookings Institution Press.

van Creveld, M. (1991) *The Transformation of War*. New York: The Free Press.

van der Haer, R. et al. (2011) Create Compliance and Cohesion – How Rebel Organizations Manage to Survive. *Small Wars & Insurgencies*. 22 (3). pp. 415–34.

Weber, M. (1978) *Economy and Society*. Berkeley: University of California Press.

Weinstein, J. & Humphreys, M. (2003) What the Fighters Say: A Survey of Ex-Combatants in Sierra Leone. June-August 2003. Online publication. Available at <http://www.columbia.edu/~mh2245/SL.htm>.

————— & Humphreys, M. (2006) Handling and Manhandling Civilians in Civil War: Determinants of the Strategies of Warring Factions. *American Political Science Review*. 100 (3). pp. 429–47.

————— (2007) *Inside Rebellion: The Politics of Insurgent Violence*. Cambridge: Cambridge University Press.

Willmott, W. (1981) Analytical Errors of the Kampuchean Communist Party. *Pacific Affairs*. 54 (2). pp. 209–27.

Wimmer, A. (2002) *Nationalist Exclusion and Ethnic Conflict: Shadows of Modernity*. Cambridge: Cambridge University Press.

Wittgenstein, L. (2009) *Philosophical Investigations*. Hoboken: John Wiley & Sons.

Wood, E. (2003) *Insurgent Collective Action and Civil War in El Salvador*. Cambridge: Cambridge University Press.

Interviews Cited in the Text

Conducted by Koy-Try Teng and the author. Recordings all in possession of author.

General Staff KPNLF (GS-KP1) Written Interview in English, E-Mail Exchange. September, 2012.

Brigade Commander KPNLF (BC-KP1) Phnom Penh. March 6, 2012.

————— (BC-KP2) Phnom Penh. May 24, 2011.

————— (BC-KP3) Borvel. March 4. 2012.

————— (BC-KP4) Kandal. July 11, 2011.

————— (BC-KP5) Svay Sisophon. June 13, 2011.

Brigade Commander Funcinpec (BC-F1) Original in English. Phnom Penh. May 20, 2011.

————— (BC-F2) Phnom Penh. February 22, 2012.

————— (BC-F3) Phnom Penh. March 20, 2012.

————— (BC-F4) Phnom Penh. October 11, 2012.

————— (BC-F5) Phnom Penh. October 2, 2012.

Brigade Commander Khmer Rouge (BC-KR1) Anlong Veng. May 13, 2012.

Brigade Commander Advisor Funcinpec (AB-F) Mongkul Borei. March 1, 2012.

Regimental Commander KPNLF (RC-KP1) Original in English. Phnom Penh. June 2, 2011.

————— (RC-KP2) Phnom Penh. March 20, 2012.

Regimental Commander Funcinpec (RC-F1) Phnom Penh. February 20, 2012.

————— (RC-F2) Pailin. June 6, 2011.

————— (RC-F3) Battambang. March 3, 2012.

Regimental Commander Khmer Rouge (RC-KR1) Anlong Veng. March 14, 2012.
—— (RC-KR2) Pailin. March 2, 2012.
—— (RC-KR3) Borvel. March 4, 2012.
—— (RC-KR4) Anlong Veng. March 14, 2012.
—— (RC-KR5) Pailin. March 2, 2012.
Battalion Commander KPNLF (BA-KP1) Svay Sisophon. June 12, 2011.
—— (BA-KP2) Kien Svay. March 18, 2012.
—— (BA-KP3) Borvel. March 4, 2012.
Battalion Commander Funcinpec (BA-F1) Phnom Penh. March 22, 2012.
—— (BA-F2) Battambang. March 3, 2012.
—— (BA-F3) Phnom Penh. February 24, 2012.
—— (BA-F4) Phnom Penh. March 22, 2012.
Battalion Commander Khmer Rouge (BA-KR1) Samlaut. June 5, 2011.
—— (BA-KR2) Kirirom. March 19, 2012.
Company Commander Khmer Rouge (CC-KR1) Pailin. March 2, 2012.
—— (CC-KR2) Kirirom. March 19, 2012.
Company Commander KPNLF (CC-KP1) Phnom Penh. July 1, 2011.
Company Commander Funcinpec (CC-F1) Pailin. March 3, 2012.
—— (CC-F2) Phnom Penh. March 21, 2012.
Section Commander Funcinpec (SC-F1) Svay Sisophon. June 11, 2011.
—— (SC-F2) Phnom Penh. March 22, 2012.
Section Commander KPNLF (SC-KP1) Camp Village. Battambang. March 11, 2012.
Group Commander KPNLF (GC-KP1) Phnom Penh. July 1, 2011.
—— (GC-KP2) Camp Village. Battambang. March 12, 2012.
Brigade Commander's Guard KPNLF (BG-KP1) Svay Sisophon. June 13, 2011.
Brigade Commander's Guard Funcinpec (BG-F1) Samraong. July 16, 2011.
Rank-and-file Khmer Rouge (RF-KR1) Pailin. March 2, 2012.
—— (RF-KR2) Pailin. March 2, 2012.
—— (RF-KR3) Pailin. March 2, 2012.
Rank-and-File KPNLF (RF-KP1) Svay Sisophon. June 12, 2011.
—— (RF-KP2) Camp Village. Battambang, March 12, 2012.
Rank-and-File Funcinpec (RF-F1) Oudung. March 23, 2012.
Rank-and-File Multiple Factions (RF-M1) Phnom Penh. July 1, 2011.
—— (RF-M2) Samraong. July 15, 2011.
Camp Leader Funcinpec (CL-F1) Surin. Thailand, June 9, 2011.
Intelligence Officer KPNLF (I-KP1) Original in English. Phnom Penh. March 21, 2012.
US Embassy Personnel (E-US1) Original in English. Phnom Penh. May 20, 2011.
Member Buddhist Association (MBA) Phnom Penh and Siem Reap. March 6 & 16, 2012.
Head Monk Buddhist Association (HM) Siem Reap. February 28, 2012.
Son Soubert. Original in English. Phnom Penh. May 19, 2011.

Documents from the 'Documentation Center of Cambodia' (DC-Cam)

Phnom Penh home broadcast (D30277) The Need to Distinguish between Patriotism and Treason [10 April 1978]. Translation in: *Searching for the Truth*. Fourth Quarter 2008. pp. 17–19.

ANON. (D34037) Relief Bodies to Cut Food to Thai Border. July 3, 1980.

ANON. (D21611) [Talking about weapon equipment]. Translated by Koy-Try Teng.

ANON. (D00591) The Party's 4-year plan to establish Socialism in all fields, 1977–1980.

ANON. (D00389) [Political lesson]. Translated by Koy-Try Teng.

ANON. (D00382) Twelve kinds of soldiers' morality. Translated by Koy-Try Teng.

Interview with Van Rith (2003) Khpop commune, S'ang district, Kandal province. February 20, 2003. Youk Chhang.

Interview with Lat Suoy (2011) Interview Series. No. 3. Interview by Long Dany. Translation by Chy Terith. May 18, 2011.

Documents from 'The Vietnam Center and Archive', Lubbock, Texas, US

(2430808004) Dien Del military of the Khmer People National Liberation Front, date unknown.

(3671302006) Interview of General Dien Del, August 22, 1994.

(3671302007) Interview of General Teap Ben, August 22, 1994.

(3671302008) Interview of General Ea Chuor Kimmeng, August 22, 1994.

(3671302009) Interview of Lieutenant Chhim Omyon, August 24, 1994.

Index